TALL BUILDINGS: STRUCTURAL SYSTEMS AND AERODYNAMIC FORM

The structural challenges of building 800 metres into the sky are substantial, and include several factors which do not affect low-rise construction. This book focusses on these areas specifically to provide the architectural and structural knowledge which must be taken into account in order to design tall buildings successfully. In presenting examples of steel, reinforced concrete and composite structural systems for such buildings, it is shown that wind load has a very important effect on the architectural and structural design. The aerodynamic approach to tall buildings is considered in this context, as is earthquake induced lateral loading.

Case studies of some of the world's most iconic buildings, illustrated in full colour, will bring to life the design challenges which they presented to architects and structural engineers. The Empire State Building, the Burj Khalifa, the Taipei 101 and the Pirelli Building are just a few examples of the buildings whose real-life specifications are used to explain and illustrate core design principles, and their subsequent effect on the finished structure.

Mehmet Halis Günel received his high school education in Ankara, and studied civil engineering at the Middle East Technical University (METU). After graduating in 1982, he continued graduate work at the same university, completing his MSc and PhD degrees in structures in 1984 and 1995 respectively. He worked as an Assistant in the Structural Mechanics Division of the Civil Engineering Department at METU from 1982–1984, and 1986–1989. He also served as Project Manager at Prokon Consultants Company in Ankara from 1989–1991. Since that time, he has been teaching in the Architecture Department at METU and is currently an Associate Professor. His technical interests comprise reinforced concrete, tall buildings, prefabrication and structural design in architecture.

Hüseyin Emre Ilgin studied architecture at METU and after graduating in 2003 he continued graduate work at the same university. He completed his MSc degree and PhD qualifications in building science and architecture in 2006 and 2010 respectively. He has worked in several architectural offices, at Cankaya University as a Scholarship Assistant, and at METU as a part-time instructor. Currently he is working as an Architect at the Ministry of Health in Ankara.

TALL BUILDINGS

Structural Systems and Aerodynamic Form

Mehmet Halis Günel and Hüseyin Emre Ilgin

Routledge
Taylor & Francis Group
LONDON AND NEW YORK

First published 2014
by Routledge

2 Park Square, Milton Park, Abingdon, Oxfordshire OX14 4RN
52 Vanderbilt Avenue, New York, NY 10017

Routledge is an imprint of the Taylor & Francis Group, an informa business

First issued in hardback 2019

British Library Cataloguing in Publication Data
A catalogue record for this book is available from the British Library

Library of Congress Cataloging-in-Publication Data
Günel, Halis.
Tall buildings : structural systems and aerodynamic form / Halis Günel, Hüseyin Emre Ilgin.
pages cm
Includes bibliographical references and index.
1. Buildings--Aerodynamics. 2. Structural frames. 3. Skyscrapers.
4. Lateral loads. 5. Wind-pressure. 6. Wind resistant design.
7. Building, Iron and steel. I. Ilgin, Hüseyin Emre. II. Title.
TH891.G87 2014
690'.21--dc23
2013043772

ISBN13: 978-1-138-02177-8 (hbk)

Typeset in Optima 10/12.5 pt by Fakenham Prepress Solutions, Fakenham, Norfolk NR21 8NN

CONTENTS

ACKNOWLEDGEMENTS

Special thanks are due to Professor Antony Wood (Executive Director, CTBUH), and to Marshall Gerometta (CTBUH), Nigel Young/Foster and Partners, and Niels Jakob Darger for their friendly interest and for providing illustrative material for this book. The authors wish to express their deepest gratitude to Professor S. Tanvir Wasti (Department of Civil Engineering, METU) for having perused the manuscript and for offering helpful suggestions.

The authors also wish to express their sincere thanks to the following:

Professor Enis Kortan (Department of Architecture, METU) for preparing the case study for the Pirelli Building;

Professor Ali Murat Tanyer (Department of Architecture, METU) for preparing the case study for the Commerzbank Tower and for his help and advice in the overall preparation of this book;

Banu Nur Sengun Dogan (Architect, METU) for preparing the case study for the Bahrain World Trade Center;

Ilkay Guryay (Architect, METU) for preparing the case study for the New York Times Tower;

and to Sinan Bilgen (Architect, METU) for preparing the case study for the Eureka Tower.

For their positive contributions in the assessment as well as production of this book for publishing, the professional efforts of Brian Guerin, Alice Aldous and Alanna Donaldson at Routledge have been vitally important, and deserve grateful recognition.

Last but not least, Leo Gough, M.Phil. (Oxon.) has to be acknowledged for his invaluable assistance in reading, and editing the manuscript. The authors highly appreciate his generous assistance.

M. Halis Günel
H. Emre Ilgin

Department of Architecture
METU
Ankara, Turkey
October, 2013

ILLUSTRATION CREDITS

The authors owe special thanks to the following people and organisations for courteously granting permission to include their illustrations in this book.

Adrian Peret: adrian.peret@gmail.com: Al Faisaliah Center, Burj Khalifa
Antony Wood/CTBUH: Chrysler Building, Hearst Magazine Tower, Empire State Building, Lake Point Tower, New York Times Tower, Park Row Building, Seagram Building, Marina City Towers, Petronas Twin Towers, Two International Finance Centre, Willis Tower, Woolworth Building
Bunyamin Bilir: CCTV Headquarters
CTBUH (Antony Wood, Jan Klerks): Aspire Tower
CTBUH (Antony Wood, Marshall Gerometta): 780 Third Avenue Building
David Randerson: Eureka Tower
Dbox Inc. (Carolina Botero/Oppenheim Architecture + Design): COR Building
Derek Forbes: Shun Hing Square, The Center
Fiona Spalding-Smith (Greg Heal / B+H Architects): First Canadian Centre
Marshall Gerometta / CTBUH: 311 South Wacker Drive, Chase Tower, The Plaza on Dewitt, John Hancock Center, Olympia Centre, One Magnificent Mile, Onterie Center
MIDMAC Contracting Co. W.L.L. (Omar Mousfi): Aspire Tower
Niels Jakob Darger: Cheung Kong Centre, Jin Mao Building, Shanghai World Financial Center, Transamerica Pyramid, Torre Agbar, Two International Finance Centre, World Tower
Nigel Young/Foster + Partners (Kathryn Tollervey / Foster + Partners): 30 St Mary Axe
PERI GmbH (Christina Hornung): Torre Agbar, HSB Turning Torso
Reiser + Umemoto, RUR Architecture, PC (Erin Kelly): O-14
RWDI (Marc Bubel): Taipei 101 wind tunnel test model
RWDI (Marc Bubel, Candice DCruz / EMAAR): Burj Khalifa wind tunnel test model
Santiago Calatrava / Samark Architecture & Design (Peter Mattsson): HSB Turning Torso
Sun Hung Kai Properties (Filipe Leung): International Commerce Centre (ICC)
Wilfried Blümler: Jin Mao Building
WS Atkins & Partners (William Grime): Bahrain World Trade Center
Abbas Riazibeidokhti: 30 St Mary Axe, **Ilkay Guryay:** New York Times Tower
Ozgur Ozturk: Hearst Magazine Tower, and **Ramazan Sari:** Zifeng Tower (Drawings created in the graduate course "Arch 536 Studies on Tall Buildings" conducted by M. Halis Gunel, H. Emre Ilgin and Ali Murat Tanyer at METU).

PREFACE

The aim of this book is to provide basic architectural and structural knowledge about the design of tall buildings. In presenting examples of the steel, reinforced concrete and composite structural systems for such buildings, it is argued that wind load has a very important effect on the architectural and structural design. The aerodynamic approach to tall buildings is considered in this context. The main readership of the book is intended to be architects, structural engineers, and their trainees. In addition, the book has been written to be accessible, as far as possible, to general readers interested in tall buildings by using plain language.

Wind and earthquake induced lateral loads have an influential role in the architectural and structural design of tall buildings. In particular, architectural design plays a large part in the precautions that can be taken to resist wind load. The aerodynamic efficiency of the building form – implicitly including the architectural concerns – and the selection of the structural system significantly affect the resistance offered by a building against lateral loads. The design of tall buildings necessitates that architects have a basic understanding of structural systems and aerodynamic forms of buildings, and that during the design process they work together with experts in other relevant fields, especially with regard to the structures and aerodynamics. Otherwise, it is possible that structural and aerodynamic solutions produced after the completion of the architectural design may be economically costly or even impossible to implement.

The book outlines the essential information that architects and structural engineers need in order to design tall buildings. In the first chapter, tall buildings are defined and their historical development is discussed; in the second chapter, wind and earthquake induced lateral loads on tall buildings are examined; in the third chapter, the structural systems of tall buildings are considered; in the fourth chapter, case-studies of a number of well-known tall buildings are presented; in the fifth chapter, the effect of wind on tall buildings is assessed; and in the final chapter, design approaches to resist wind effects on tall buildings are reviewed.

INTRODUCTION

Throughout history, human beings have built tall monumental structures such as temples, pyramids and cathedrals to honour their gods. Human beings have always been struggling to push the limits of nature in their age-old quest for height, from the legendary Tower of Babel in antiquity, purportedly designed with the aim of reaching heaven, to today's tallest building. Today's skyscrapers are monumental buildings too, and are built as symbols of power, wealth and prestige.

At the beginning of the twentieth century, tall buildings were generally designed as offices, and achieved an important position as a "distinguished space" in the history of American urban architecture. These buildings emerged as a response to the rapidly growing urban population, with the aim of meeting the demand for office units to be positioned as closely as possible to one another. Architects' creative approaches in their designs for tall buildings, the shortage and high cost of urban land, the desire to prevent disorderly urban expansion, the effort to create a skyline concept, and factors such as concerns for a cultural identity and for prestige have driven the increase in the height of buildings.

Today it is almost impossible to imagine a major city without tall buildings. As the most important symbols of today's cities, tall buildings have become a source of faith in technology and national pride, and have changed the concept of the modern city along with its scale and appearance. Despite the fact that tall buildings have moved city life away from the human scale, in general it is accepted that these buildings are an inevitable feature of urban development.

In the past, the forms used in design were restricted but currently freedom in the design of tall buildings has significantly increased, along with a contemporary widening of the form spectrum in design. Tall buildings today, designed with the aid of advanced computer technologies, are built with exceedingly daring architectural and structural designs that are almost never found in their predecessors.

The most important factors enabling the construction of tall buildings are developments and innovations in the following areas: materials, construction techniques, operating (mechanical) systems, structural systems and analysis, but at the same time, the increase in the height of buildings makes them vulnerable to wind and earthquake induced lateral loads.

Essentially, in tall building design, which aims to respond to the needs of the occupants, in addition to structural safety, standards of occupancy comfort (serviceability) are also among the foremost design inputs. Excessive building sway due to wind can cause damage to non-structural elements, the breakage of windows, the shortening of fatigue life, the malfunction of elevators and other mechanical equipment, and damage to, or even the failure of, a structural system. In this regard, wind induced building sway affects both the structural safety and the serviceability of a building, and is thus a critical variable in the design of tall buildings. As a result, building sway becomes a serious problem for designers as much as for occupants, and during windstorms it is necessary to keep it within acceptable limits, especially to reduce the discomfort felt by occupants on the top floors to a minimum and to prevent the negative outcomes discussed above.

Much research has been done with the aim of improving the performance of buildings against wind loads. With the aim of controlling wind induced building sway and fully ensuring the functional performance of tall and slender buildings, the following approaches are used:

1. *Architectural design approach:* aerodynamic-based and structure-based design.
2. *Structural design approach:* shear-frame, mega column, mega core, outriggered frame and tube systems.
3. *Mechanical design approach:* auxiliary damping systems.

The design of tall buildings is a complex subject that requires interdisciplinary collaboration and teamwork at an advanced level, and architects need to be aware of this fact. In tall building design, the increase in the dimensions of structural elements, differences in the structural system and the operating (mechanical) systems, and the aerodynamic/structure-based building form, are foremost in their effect on architectural and structural design. When architects take into account basic aerodynamic principles, the reduction of wind induced building sway plays an important role. Thus it reduces costs by lowering the demands on the structural system and the auxiliary damping system substantially, and at the same time it reduces to a minimum the modifications that may be required after wind tunnel testing. In considering the critical role played by wind and earthquake induced lateral loads in design decisions for tall buildings, it is vital to consider architectural design alongside structural design and aerodynamic design approaches. As a result, it becomes difficult to speak of architects as having a satisfactory degree of freedom in designing tall buildings. In this context, according to the authors, "Skyscraper design is inevitably the output of interdisciplinary teamwork, led by the architect, who takes structural and aerodynamic issues into consideration while struggling not to sacrifice the architectural design."

The altitudes that have been possible to reach, from the late nineteenth century to the present, are as follows: in 1885 the first acknowledged skyscraper, the Home Insurance Building (Chicago) reached 55 m; in 1931, the Empire State Building (New York) reached 381 m; and in 2010, the Burj Khalifa (Dubai) reached 828 m. Today, the race for height continues at an accelerating pace, thanks to innovations and advances in structural analysis and design, and improvements in high-strength materials and construction technology. This race strains the boundaries, engendering new architectural and engineering problems that need to be solved. Thus architects have design freedom, but only to the limits of what is possible in engineering and technology. In order to solve the problems created by the increasing heights of buildings, an architect has an aesthetic struggle with limited design freedom, since the structural strength of a building and the loads to which it is exposed affect the choice of form to a great degree, and other engineering disciplines, especially structural and aerodynamic disciplines, also contribute to the design during collaborative teamwork.

In skyscraper design, from the beginning, notions of "uniqueness" and "being a symbol" have usually been very important. Most skyscrapers make their mark and are thought to be successful to the extent to which these notions are reflected in the design, along with the height of the building.

The following are mentioned as prominent designs:

- World Trade Center Twin Towers (New York, 1972), facade
- Chrysler Building (New York, 1930) and the Empire State Building (New York, 1931), sculptured building top
- John Hancock Center (Chicago, 1969), tapered (truncated pyramid) form and a structural expressionist trussed-tube structure
- Willis Tower (Chicago, 1974), structural expressionist bundled-tube form
- Taipei 101 (Taipei, 2004), bamboo form
- Burj Khalifa (Dubai, 2010), flower form
- HSB Turning Torso (Malmö, 2005) and the Chicago Spire (Chicago, under construction), twisted form.

Thus, in the continual race for height, city skylines are being shaped by unconventional forms.

1

TALL BUILDINGS

"Tall building", "high-rise building" and "skyscraper" are difficult to define and distinguish solely from a dimensional perspective because height is a relative matter that changes according to time and place. While these terms all refer to the notion of very tall buildings, the term "skyscraper" is the most forceful. The term "high-rise building" has been recognised as a building type since the late nineteenth century, while the history of the term "tall building" is very much older than that of the term "high-rise building". As for the use of the term "skyscraper" for some tall/high-rise buildings reflecting social amazement and exaggeration, it first began in connection with the 12-storey Home Insurance Building, built in Chicago towards the end of the nineteenth century (Harbert, 2002; Peet, 2011).

1.1 Definition

There is no general consensus on the height or number of storeys above which buildings should be classified as tall buildings or skyscrapers. The architectural/structural height of a building is measured from the open-air pedestrian entrance to the top of the building, ignoring antennae and flagpoles. According to the CTBUH[1] (Council on Tall Buildings and Urban Habitat), buildings of 14 storeys or 50 metres' height and above could be considered as "tall buildings"; buildings of 300 metres' and 600 metres' height and above are classified as "supertall buildings" and "megatall buildings" respectively. The CTBUH measures the "height to architectural top" from the level of the lowest "significant open-air pedestrian entrance" to the architectural top of the building, including spires, but not including antennae, signage, flag poles or other functional-technical equipment. In this book, this height measurement is used for the "architectural height" of the buildings.

1 CTBUH, Council on Tall Buildings and Urban Habitat, Illinois Institute of Technology, S.R. Crown Hall, 3360 South State Street, Chicago, Illinois, USA, www.ctbuh.org.

According to the Emporis Standards, buildings of 12 storeys or 35 metres' height and above, and multi-storey buildings of more than 100 metres' height, are classified as "high-rise buildings" and "skyscrapers" respectively (Emporis Data Standards ESN 18727, ESN 24419).[2]

According to Ali and Armstrong, the authors of *Architecture of Tall Buildings* (1995),

> the tall building can be described as a multistorey building generally constructed using a structural frame, provided with high-speed elevators, and combining extraordinary height with ordinary room spaces such as could be found in low-buildings. In aggregate, it is a physical, economic, and technological expression of the city's power base, representing its private and public investments.

Beedle (1971) defines a "tall building" as a multi-storey building that requires additional construction techniques because of its extraordinary height.

Tall buildings are defined: by structural designers as buildings that require an unusual structural system and where wind loads are prominent in analysis and design; by architectural designers as buildings requiring interdisciplinary work in particular with structural designers, and with experts in the fields of aerodynamics, mechanics and urban planning that affect design and use; and by civil engineers as buildings needing unusual and sophisticated construction techniques.

The first use of the word "skyscraper" in the sense of "tall building" was in an article published in 1883 in the journal *American Architect*, appearing as "America needs tall buildings; it needs skyscrapers" (Giblin, 1981). While Ada Louise Huxtable (1984) emphasises that tall buildings are symbols of our age and that the words "skyscraper" and "twentieth century" have an equivalent meaning, César Pelli (1982) defines a skyscraper as a supertall building and highlights the word "super" within this definition as changing according to time and place. Structures such as the Eiffel Tower (Paris, 1889) cannot be classified as skyscrapers because of the lack of a habitable interior space.

In the view of the authors of this book, "tall building or high-rise building" is a local concept and "skyscraper or supertall building" is a global concept. To be able to define a tall building as a skyscraper or supertall building, it is not sufficient for it only to be tall in its own region; it is necessary for it to be recognised around the world as a skyscraper or supertall building. In this context skyscraper or supertall building is distinguished as being higher than tall or high-rise building.

1.2 Emergence and historical development

> No other symbols of the modern era are more convincing than the gravity defying, vertical shafts of steel, glass, and concrete that are called "skyscrapers."
>
> (Harbert, 2002)

2 Emporis, Emporis Corporation, A Global Building Information Company, Theodor-Heuss-Allee 2, 60486 Frankfurt, Germany, www.emporis.com.

Like the Greek temples or the Gothic cathedrals that were the foremost building types of their own ages, skyscrapers have become iconic structures of industrial societies. These structures are an architectural response to the human instincts, egos and rivalries that always create an urge to build higher, and to the economic needs brought about by intense urbanisation.

Architects make a contribution to the social and economic changes of the age, reflecting the environment they live in with their designs and creating a development/evolution by developing new styles or building types. In addition, underlying the first appearances of skyscrapers in Chicago was a social transformation triggered by the economic boom of that era and by the increase in value of urban building plots. The concentrated demand for increasing incorporation in city centres, together with the intensification of business activity and the rise in the values of capitalism, necessitated the creation of a new, unusually high building type which had the large spaces that could meet these demands – and many such buildings were produced using extraordinary forms and techniques.

In the masonry construction technique that was employed before the development of rigid frame (beam-column framing) systems, load-bearing masonry walls were used structurally, which, although they had high levels of fire resistance, reduced the net usable area because of their excess dead loads and wide cross-sections. The 64 metres, attained towards the end of the nineteenth century by the 17-storey Monadnock Building (Chicago, 1891), is the highest point that this construction technique was able to reach (Figure 1.1). The structure used 2.13 m thick load-bearing masonry walls at the ground floor, and was the last building to be built in the city using this technique.

FIGURE 1.1 Monadnock Building, Chicago, USA, 1891

The advance in construction technology has played a much more important role in the development of tall buildings than in the case of other types of structure. At the end of the nineteenth century, beginning with the discovery of the elevator for the vertical transportation system, and structural metal (cast iron which was soon replaced by steel) beam-column framing system, the construction of tall buildings commenced as an American building type owing to innovations and developments in new structural systems, high-strength concrete, foundation systems and mechanical systems; this continues to drive the race for height in skyscrapers that is spreading across the world.

The Home Insurance Building (Chicago, 1885) (Figure 1.2), designed by engineer William Le Baron Jenney with 12 storeys (2 storeys were added later), is recognised as being the first skyscraper. The use of a structural frame in the building won it the title of the first skyscraper, marking a new epoch in the construction of tall buildings, and it became a model for later tall building designs.

After the Home Insurance Building (Chicago) in 1885 at 55 m, the race to construct the world's tallest building continued with:

- World Building (New York) in 1890 at 94 m
- Manhattan Life Insurance Building (New York) in 1894 at 106 m
- Park Row Building (New York) in 1899 at 119 m (Figure 1.3)
- Singer Building (New York) in 1908 at 187 m (Figure 1.4)
- Metropolitan Life Tower (New York) in 1909 at 213 m
- Woolworth Building (New York) in 1913 at 241 m (Figure 1.5)
- Trump Building (New York) in 1930 at 283 m (Figure 1.6)
- Chrysler Building (New York) in 1930 at 319 m (Figure 3.16)
- Empire State Building (New York) in 1931 at 381 m (Figure 3.17)
- One World Trade Center (WTC I) (New York) in 1972 at 417 m (Figure 3.55)
- Two World Trade Center (WTC II) (New York) in 1973 at 415 m (Figure 3.55)
- Willis Tower (Chicago) in 1974 at 442 m (Figure 3.73)
- Petronas Twin Towers (Kuala Lumpur) in 1998 at 452 m (Figure 3.31)
- Taipei 101 (Taipei) in 2004 at 508 m (Figure 3.36)
- Burj Khalifa (Dubai) in 2010 at 828 m (Figure 3.30)

and when 800 m was passed at the beginning of the 2000s, heights have been reached that could not have even been dreamed of in engineer William Le Baron Jenney's time. In other words, while 10-storey buildings were classified as skyscrapers in the 1890s, about 40 years later the Empire State Building (New York, 1931) exceeded 100 storeys, and about 100 years later the Burj Khalifa (Dubai, 2010) exceeded 150 storeys. Skyscrapers, which were thought previously to be exclusively a North American urban phenomenon, have today entered the skylines of almost all major cities, especially in Asia.

According to the CTBUH, while in 1930, 99 per cent of the world's 100 tallest buildings were in North America, 51 per cent of which were in New York, in 2010 these proportions were 29 per cent and 7 per cent respectively. In 1930, 96 per cent of the world's 100 tallest buildings had steel and 4 per cent had reinforced concrete and composite structural systems, but in 2010 these figures had become 21 per cent and 79 per cent respectively.

While in 1990, 9 out of 10 of the world's tallest buildings were in North America, this figure had dropped to 4 in 2000 and to 1 in 2011. In 1990, 8 out of the 10 tallest buildings had steel structural systems, and this figure had fallen in 2000 and 2011 to 4 and 1 respectively (Tables 1.1–1.3).

FIGURE 1.2 Home Insurance Building, Chicago, USA, 1885

FIGURE 1.3 Park Row Building, New York, USA, 1899
(courtesy of Antony Wood/CTBUH)

FIGURE 1.4 Singer Building, New York, USA, 1908

FIGURE 1.5 Woolworth Building, New York, USA, 1913 *(photo on right courtesy of Antony Wood / CTBUH)*

FIGURE 1.6 The Trump Building, New York, USA, 1930

TABLE 1.1 The world's ten tallest buildings in 1990 (CTBUH, October, 2009)

Rank	*Building name*	*Location*	*Completion*	*Height (m)*	*Structural material*
1	Sears Tower (currently Willis Tower)	Chicago	1974	442	Steel
2	One World Trade Center (WTC I)	New York	1972	417	Steel
3	Two World Trade Center (WTC II)	New York	1973	415	Steel
4	Empire State Building	New York	1931	381	Steel
5	Bank of China Tower	Hong Kong	1990	367	Composite
6	Aon Center (formerly Amoco Building)	Chicago	1973	346	Steel
7	John Hancock Center	Chicago	1969	344	Steel
8	Chrysler Building	New York	1930	319	Steel
9	U.S. Bank Tower (formerly Library Tower)	Los Angeles	1990	310	Steel
10	Franklin Center-North Tower (formerly AT&T Corporate Center)	Chicago	1989	307	Composite

TABLE 1.2 The world's ten tallest buildings in 2000 (CTBUH, October, 2009)

Rank	*Building name*	*Location*	*Completion*	*Height (m)*	*Structural material*
1	Petronas Tower 1	Kuala Lumpur	1998	452	R/C
2	Petronas Tower 2	Kuala Lumpur	1998	452	R/C
3	Sears Tower (currently Willis Tower)	Chicago	1974	442	Steel
4	Jin Mao Building	Shanghai	1999	421	Composite
5	One World Trade Center	New York	1972	417	Steel
6	Two World Trade Center	New York	1973	415	Steel
7	CITIC Plaza	Guangzhou	1996	390	Composite
8	Shun Hing Square	Shenzhen	1996	384	Composite
9	Empire State Building	New York	1931	381	Steel
10	Central Plaza	Hong Kong	1992	374	Composite

TABLE 1.3 The world's ten tallest buildings in 2011 (CTBUH, December, 2011)

Rank	*Building name*	*Location*	*Completion*	*Height (m)*	*Structural material*
1	Burj Khalifa (formerly Burj Dubai)	Dubai	2010	828	R/C
2	Taipei 101	Taipei	2004	508	Composite
3	Shanghai World Financial Center	Shanghai	2008	492	Composite
4	International Commerce Centre (ICC)	Hong Kong	2010	484	Composite
5	Petronas Tower 1	Kuala Lumpur	1998	452	R/C
6	Petronas Tower 2	Kuala Lumpur	1998	452	R/C
7	Zifeng Tower (formerly Nanjing Greenland Financial Center)	Nanjing	2010	450	Composite
8	Willis Tower (formerly Sears Tower)	Chicago	1974	442	Steel
9	Kingkey 100	Shenzhen	2011	442	Composite
10	Guangzhou International Finance Center	Guangzhou	2010	439	Composite

2

LATERAL LOADS AFFECTING TALL BUILDINGS

From the structural design point of view, tall (high-rise) buildings, because of their extraordinary height, show a greater sensitivity to wind and earthquake induced lateral loads than low-rise buildings. Estimating those lateral loads which play an important role in the design of tall buildings is more difficult than estimating vertical loads.

Earthquake loads increase according to the building weight, and wind loads increase according to the building height. For this reason, wind loads, while they are generally an unimportant issue in the design of structural systems for low- and mid-rise buildings, play a decisive role in that of tall buildings, and can even be a cause of large lateral drift (sway) that is more critical than that from earthquake loads. Consequently, the occupancy comfort takes prominence in the design of structural systems in tall buildings, and it is necessary to limit the building sway. In tall buildings, which can be described as vertical cantilever beams, the maximum lateral top drift caused by lateral loads is expected to be approximately 1/500 of the building height (structural height), according to Bennett (1995) and Taranath (1998), and in limits ranging from 1.5/1000 to 3/1000 according to Smith and Coull (1991). In this context, the drift index is defined as the ratio of the maximum lateral top displacement of the building to the building height (Δ/H); and the inter-storey drift index as the ratio of the lateral displacement of the floor relative to the floor below, to the floor-to-floor height (Δ/h). Generally in wind design of tall buildings, 1/400–500 is commonly preferred as both the drift index and the inter-storey drift index.

2.1 Wind loads

At first wind loads were ignored because the weight of the construction materials and structural systems used in the first skyscrapers made vertical loads more critical than lateral loads, but over time wind loads became important, as the strength to weight ratio of construction materials and the ratio of floor area to structural weight in structural systems increased and the total weight and rigidity of structures decreased. The effect of wind on tall buildings is explained in Chapter 5.

Wind speed and pressure increase parabolically according to height, and therefore wind loads affecting tall buildings become important as the height of the building increases. In general, structural design begins to be controlled by wind loads in buildings of more than 40 storeys (ACI SP-97, 1989). Today, thanks to developments in structural systems and to high-strength materials, tall buildings have increased in their height to weight ratio but on the other hand reduced in stiffness compared with their precursors, and so have become greatly affected by wind. With the reduced stiffness, the sensitivity to lateral drift, and hence the sway under wind loads, increases. The sway, which cannot be observed outside the building or at the lower floors, can cause discomfort to occupants at the higher floors of a building. Architectural, structural, and mechanical design approaches (Chapter 6) are used to control lateral drift in tall buildings.

In the design of tall buildings, for buildings below 40 storeys with height to width ratio (the ratio of the structural height of a building to the narrowest structural width at the ground floor plan, also termed aspect ratio) below 6, the values predicted in the building design codes can be used to determine wind loads. Because wind loads can change quickly or even suddenly, unlike live and dead loads, in order to estimate the wind load in buildings of more than 40 storeys, or that have an aspect ratio of 6 or higher (slender and flexible buildings), or that have unusual forms, dynamic effect of the wind and dynamic building response must be taken into account. In this context, dynamic calculation methods, or else wind tunnel tests, are recommended for estimating the wind loads on such buildings (Section 5.2).

2.2 Earthquake loads

Earthquakes are the propagation of energy released as seismic waves in the earth when the earth's crust cracks, or when sudden slippage occurs along the cracks as a result of the movement of the earth's tectonic plates relative to one another. With the cracking of the earth's crust, faults develop. Over time, an accumulation of stress in the faults results in sudden slippage and the release of energy. The propagation of waves of energy, formed as a result of seismic movement in the earth's crust, acts upon the building foundations and becomes the earthquake load of the building. In determining earthquake loads, the characteristics of the structure and records of previous earthquakes have great importance. Compared with wind loads, earthquake loads are more intense but of shorter duration.

Earthquakes can occur almost anywhere, and considering that low, medium and high severity earthquakes may occur during the life of a structure located in an active earthquake zone, it is necessary to understand very well the behaviour of a structure during an earthquake in order to prevent the disastrous collapses that can occur.

An earthquake's effect or power is measured by the "earthquake's intensity" or "earthquake's magnitude". Accounting for the effects upon living creatures, structures and the environment in the measurement of an earthquake gives the "intensity" of the earthquake, while using earthquake seismographs (seismometers) to measure the energy released at the centre of an earthquake gives the "magnitude" of the earthquake. The intensity of an earthquake indicates its effect in any given region. The magnitude of an earthquake gives information on its intensity at its centre (epicentre). While the measure of magnitude gives only a single value for the magnitude of an earthquake, the measure of intensity gives different intensity values in different regions. The "magnitude" of earthquakes is indicated by the *Richter scale* and their "intensity" is indicated by the *Mercalli scale*.

The lateral inertia forces on a structure created by an earthquake are functions of:

- the magnitude and duration of the earthquake
- the distance of the structure from the centre of the earthquake (epicentre) and
- the mass of the structure, the structural system and the soil-structure interaction.

The magnitude of the lateral force (F) on a structure formed by the effect of an earthquake depends on the structure's mass (m), the ground acceleration (a) and the structure's dynamic characteristics ($F \propto ma$) (Figure 2.1). The ground acceleration changes according to the characteristics of the earthquake and the ground. Theoretically, in the case of rigid structures and foundations, the acceleration of the structure is equal to that of the ground. In this case, according to Newton's Law, the lateral load (F) affecting a structure is equal to the mass (m) of the structure multiplied by the ground acceleration (a), ($F = ma$) (Figure 2.1a). This theoretical case does not occur in practice because every structure has certain flexibility. For a structure that deforms due to its flexibility, thus dissipating some energy, the lateral force (F) affecting the structure is less than the product of the mass of the structure and the ground acceleration ($F<ma$) (Figure 2.1b). As the height of a structure increases, the flexibility also increases and the acceleration is expected to be less than in low-rise

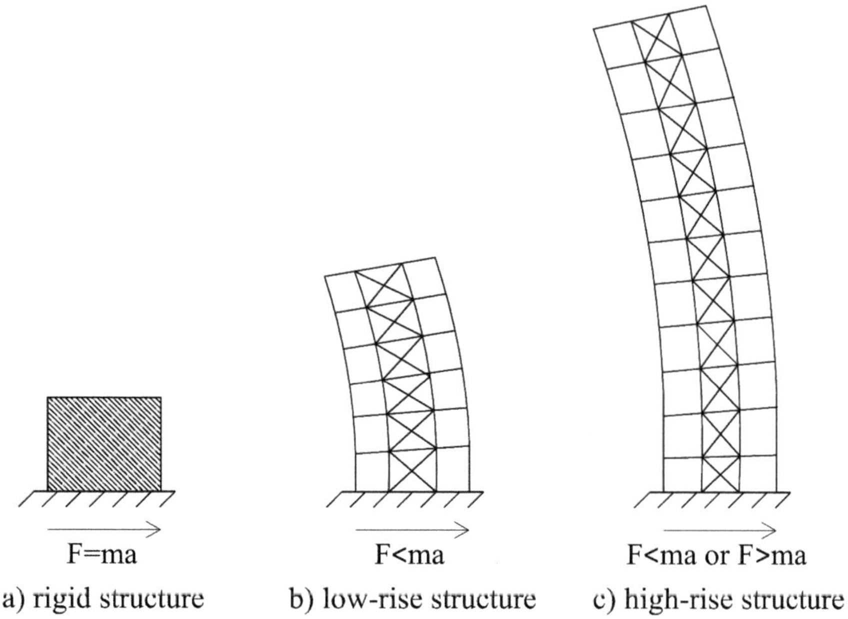

FIGURE 2.1 The behaviour of a building during an earthquake

structures ($F<ma$); however, for structures whose natural period is close to that of the seismic waves, in earthquakes of long duration, the lateral force (F) affecting the structure may be larger than the mass of the structure multiplied by the ground acceleration ($F>ma$) (Figure 2.1c). For this reason, the lateral load on a structure caused by an earthquake is a function not only of the mass of the structure and the ground acceleration, but also of the dynamic characteristics of the structure.

The general principle in earthquake-resistant design is to protect life without a collapse of the structure, even if there is damage to structural and non-structural elements of the building. Earthquake codes aim to ensure: the avoidance of damage to structural and non-structural elements of the building during earthquakes of low intensity; the limitation and reparability of the damage that may occur to structural and non-structural elements during earthquakes of medium intensity; the avoidance of the collapse of structural elements where there is limited and permanent damage, and the protection of life during earthquakes of high intensity.

3

THE STRUCTURAL SYSTEMS OF TALL BUILDINGS

Structural systems in the early twentieth century buildings were basically designed to resist vertical loads. Today, thanks to developments in this field and to high-strength materials, with the increase in the height of buildings and the decrease in their weight, wind and earthquake induced lateral loads have become the primary loads, especially in tall buildings, and have begun to pose more of a threat than before. As a result, for structural engineers, providing the strength to resist lateral loads in tall buildings, whether wind or earthquake induced, has become an essential input in the design of new structural systems.

Owing to developments in computer technology, construction materials and structural design, tall building structural systems have gone far beyond the rigid frame system of the 12-storey, 55 m high Home Insurance Building (Chicago, 1885) (Figure 1.2), recognised as the first skyscraper, and have today reached a point that could not have been dreamed of in Le Baron Jenney's time, attaining a level that has made possible the construction of buildings using outriggered frame systems, such as the 101-storey, 508 m high Taipei 101 (Taipei, 2004) (Figure 3.36), and the 163-storey, 828 m high Burj Khalifa (Dubai, 2010) (Figure 3.30).

As the height of buildings increases, the choice of structural system decreases. While the choice of structural system in low-rise buildings is considerable, the alternatives in choice of a structural system become restricted by limitations imposed by the height of buildings. Therefore, especially in tall buildings, architectural and structural design should be considered together.

Buildings can be classified on the basis of the materials used in their structural systems [structural materials of the columns, beams, shear trusses (braces), shear walls and outriggers] as:

- steel
- reinforced concrete
- composite.

Taking as a basis the columns, beams, shear trusses (braces), shear walls, and outriggers that are the elements of the main vertical and horizontal structural systems, buildings can be categorised as being reinforced concrete buildings where these elements are made of reinforced concrete, or as steel buildings where these elements are made of steel. We can define composite buildings as: those in which some structural elements are made of reinforced concrete and other structural elements are made of steel; and/or those in which some structural elements are made of both structural steel and concrete together. Floor slabs are usually made of reinforced concrete or are composite. Thus, it is general practice to use concrete/reinforced concrete in slabs. Generally, the floor slabs in steel buildings are composite, and in reinforced concrete and composite buildings they are reinforced concrete or composite. Normally, composite floor slabs are formed by applying metal deck (trapezoidal steel plate) with concrete/reinforced concrete topping. Composite floor slabs commonly support structural steel or steel trusses. According to the Council on Tall Buildings and Urban Habitat, in classifying tall buildings based on the materials used in their structural systems, the main vertical and horizontal structural elements but also the floor systems are taken into account.

The use of steel as a material for a structural system attracted attention in 1885 with the construction of the 55 m high Home Insurance Building (Chicago) (Figure1.2) and of the 300 m high Eiffel Tower (Paris) in 1889. By the end of the 1990s all the buildings achieving the title of "the world's tallest building" used steel structural systems due to the superiority of structural steel in its strength to weight ratio, the ease with which it can be transported, installed and assembled on-site, the wide range in the choice of strength and cross-sections of the elements, and advances in fire and corrosion resistance. The 442 m high Willis Tower (Chicago, 1974) (Figure 3.73), with a steel structural system, held the title of "the world's tallest building" from 1974 to 1998.

Reinforced concrete is formed by strengthening concrete with steel bars. The discovery of reinforced concrete greatly increased the importance and use of concrete in the construction industry. Architects and structural engineers used reinforced concrete to produce unusual and aesthetic building forms, thanks to its ability to be cast in any form, and to its much greater natural resistance to fire, compared with steel. In addition, compared with a steel building, a reinforced concrete building is naturally better at dampening wind induced building sway, which is one of the problems frequently encountered in tall buildings and is perceived by the building occupants. With the advances in technology, the increase in strength and developments in concrete pumping technology – the ability to pump it to high levels – reinforced concrete can now be used in all structural systems for tall buildings. The 16-storey, 65 m high Ingalls Building, built by Elzner and Anderson in Cincinnati in 1903, was the first skyscraper with a reinforced concrete structural system (Figure 3.1). In 1998 the Petronas Twin Towers, 452 m high and with a reinforced concrete structural system (Figure 3.31), took the title of "the world's tallest building" from the Willis Tower, which has a steel structural system. The Petronas Twin Towers were the first reinforced concrete buildings that gained the title of the world's tallest.

While composite structural systems, consisting of steel and reinforced concrete together, were only rarely seen in supertall buildings before 1970 (Chrysler Building

FIGURE 3.1 Ingalls Building, Cincinnati, USA, 1903

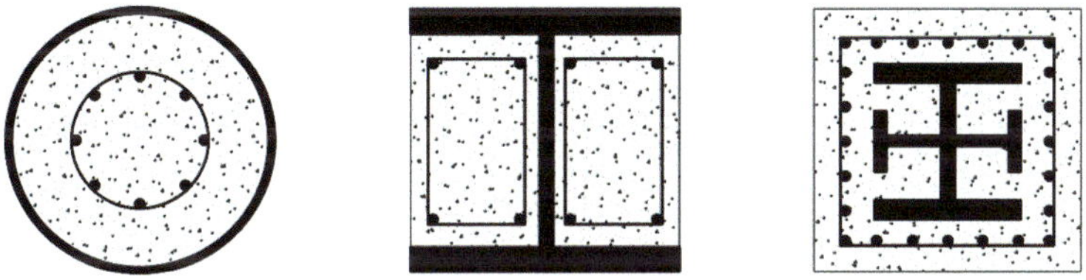

FIGURE 3.2 Composite elements by cross-section

(New York, 1930), Seagram Building (New York, 1958)), they began to be used frequently after the 1970s.

Composite buildings, consisting of structural system elements that are part reinforced concrete, part steel and/or with some elements in which both steel and reinforced concrete have been used, combine the advantages of both materials, such as the high-strength of steel, and the fire resistance and rigidity of reinforced concrete.

There are differences in the cross-sections of composite elements (Figure 3.2). Box-section structural steel elements filled with reinforced concrete (concrete infilled steel sections/steel encased concrete sections) (Figure 3.2a), structural steel elements with reinforced concrete between their flanges (Figure 3.2b) and structural steel sections encased in reinforced concrete (concrete encased steel sections) (Figure 3.2c) are all seen as elements of a composite structural system.

In 2004, the 508 m high Taipei 101 (Taipei) (Figure 3.36), with a composite structural system, took the title of "the world's tallest building" from the Petronas Towers (Figure 3.31), which has a reinforced concrete structural system. The Taipei 101 is the first composite building to gain the title of the world's tallest.

With the use of high-strength concrete (compressive strength above 30 MPa) in the 828 m high reinforced concrete Burj Khalifa (Dubai, 2010) (compressive strength of 80 MPa); 452 m high reinforced concrete the Petronas Twin Towers (Kuala Lumpur, 1998) (compressive strength of 80 MPa) and 421 m high, composite Jin Mao Building (Shanghai, 1999) (compressive strength of 52 MPa), the cross-sectional areas of structural elements have been reduced, and thus the reinforced concrete building construction has been improved and came into prominence.

When the 10 tallest buildings in 2011, based on the vertical and horizontal elements of the main structural systems are evaluated, it is evident that 1 is steel, 3 are reinforced concrete and 6 are composite, making a 90 per cent majority that are reinforced concrete or composite buildings (Burj Khalifa (Dubai, 2010) reinforced concrete, Taipei 101 (Taipei, 2004) composite, Shanghai World Financial Center (Shanghai, 2008) composite, International Commerce Centre (ICC) (Hong Kong, 2010) composite, Petronas Tower 1 and Petronas Tower 2 (Kuala Lumpur, 1998) reinforced concrete, Zifeng Tower (Nanjing, 2010) composite, Willis Tower (Chicago, 1974) steel, Kingkey 100 (Shenzhen, 2011) composite, Guangzhou International Finance Center (Guangzhou, 2010) composite).

3.1 The structural systems of tall buildings

The set of tall building structural systems has developed over time, starting with rigid frame systems, and with the addition of shear-frame, mega column (mega frame, space truss), mega core, outriggered frame, and tube systems, it has made much taller buildings possible.

Today, many tall building structural systems and classifications are discussed in the literature and used in practice (Khan, 1969; Khan, 1973; Schueller, 1977; Smith and Coull, 1991; Taranath, 1998). Steel, reinforced concrete and composite structural systems for tall buildings can be categorised by their structural behaviour under lateral loads.

Tall building structural systems:

- rigid frame systems
- flat plate/slab systems
- core systems
- shear wall systems
- shear-frame systems
    - shear trussed frame (braced frame) systems
    - shear walled frame systems
- mega column (mega frame, space truss) systems
- mega core systems
- outriggered frame systems[1]
- tube systems
    - framed-tube systems
    - trussed-tube systems
    - bundled-tube systems.

1 An "outrigger" consists of a horizontal truss or shear wall. The element, which is called an "outrigger", is a lateral extension of the core shear truss/shear wall to the perimeter columns in the form of a knee. An outriggered frame system is formed by the addition of an outrigger to a shear-frame system having a structural core (core-frame system). When an outrigger is used in a tube system, the structural system is referred to as a "tube system".

The examples of systems listed above are given in Chapter 4 and the Appendix under "Tall Building Case Studies" and "Examples of Tall Buildings and Structural Systems".

For "tall buildings" of 40 storeys and below, "rigid frame systems", "flat plate/slab systems", "core systems" and "shear wall systems" are used. For "supertall buildings" and "skyscrapers" over 40 storeys, the necessity for an economic and efficient structural system satisfying both the structural safety and serviceability (occupancy comfort) to be limited to a maximum lateral drift due to lateral loads of approximately 1/500 of the building height, reduces the choice of structural system. For this reason, for buildings of more than 40 storeys, "shear-frame systems", "mega column systems", "mega core systems", "outriggered frame systems" and "tube systems" are used.

Supertall building / skyscraper structural systems:

- shear-frame systems
    - shear trussed frame/braced frame systems
    - shear walled frame systems
- mega column (mega frame, space truss) systems
- mega core systems
- outriggered frame systems
- tube systems
    - framed-tube systems
    - trussed-tube systems
    - bundled-tube systems.

The number of floors that can be reached efficiently and economically by structural systems for tall buildings and supertall buildings/skyscrapers is shown in Table 3.1.

The structural system in supertall buildings/skyscrapers is closely connected to the form and function of the building, and thus to architectural design. The choice of structural system has an important effect on the building facade and interior use (the service core, including elevators, stairs, emergency exits and wet areas, and the net usable area other than the service core). From an economic point of view, local materials and construction techniques play a decisive role in the use of steel,

TABLE 3.1 Tall building structural systems and the number of floors they can reach

Tall building structural systems, and tentatively the number of floors they can reach efficiently and economically	*10*	*20*	*30*	*40*	*>40*
Rigid frame systems	▬	▬	▬		
Flat plate/slab systems with columns and/or shear walls	▬	▬	▬		
Core systems	▬	▬			
Shear wall systems	▬	▬	▬	▬	
Shear-frame systems (shear trussed / braced frame and shear walled frame systems)	▬	▬	▬	▬	▬
Mega column (mega frame, space truss) systems	▬	▬	▬	▬	▬
Mega core systems	▬	▬	▬	▬	▬
Outriggered frame systems	▬	▬	▬	▬	▬
Tube systems	▬	▬	▬	▬	▬

reinforced concrete or composite in the structural system. In North America, steel construction, while in the Asia Pacific (Far East), reinforced concrete or composite construction is generally used.

With the progress in technology, the increase in the strength of concrete and the ability to pump concrete to high levels, the use of reinforced concrete and composite structural systems increases day by day.

3.2 Rigid frame systems

Rigid frame systems, also called moment frame systems, are used in steel and reinforced concrete buildings. This system consists of beams and columns (Figure 3.3). A rigid frame is an unbraced frame that is capable of resisting both vertical and lateral loads by the bending of beams and columns. Stiffness of the rigid frame is provided mainly by the bending rigidity of beams and columns that have rigid connections. Rigid framing is based on the principle that beam-column connections have adequate rigidity to hold the original angles between intersecting members unchanged under the effect of both vertical and lateral loads. Thus, reinforced concrete is an ideal material for this system by virtue of its naturally monolithic behaviour, resulting with inherent rigidity at connections. For steel buildings, rigid framing is achieved by reinforcing beam-column connections.

The structural stiffness of rigid frames is directly proportional to the cross-sectional dimensions and bending rigidity of the beams and columns, and inversely proportional to their length and spacing. In this system, columns are placed in locations that least restrict architectural planning. At the same time, columns should be of sufficient length to provide minimum storey depth. To obtain effective rigid frame behaviour, it is necessary to have closely spaced columns, and for the beams connecting them to be sufficiently deep.

For buildings constructed in regions of high seismic activity in particular, the details of the connections between structural elements are very important because of the need for ductile behaviour in the rigid frame due to the large lateral drift during

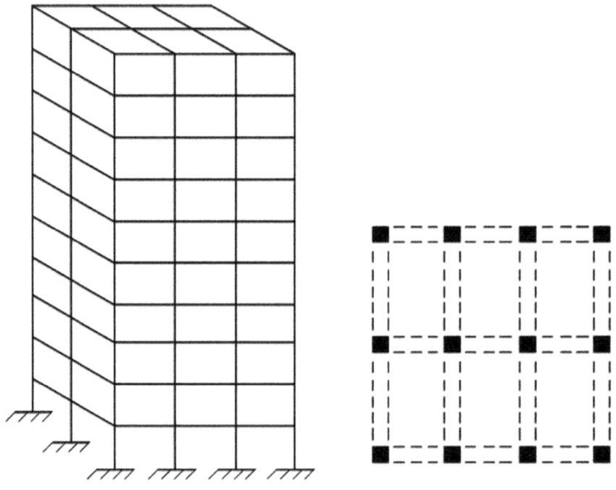

FIGURE 3.3 Rigid frame system

severe earthquakes (ductility is the ability to deform without a significant reduction in strength). In rigid frame systems ductility is achieved by the formation of plastic hinges in the columns and beams. In this way, when rigid frame systems under earthquake loads are deformed beyond their elastic limits, a large part of the energy is dissipated by the plastic hinges. Steel is a ductile material, while concrete is a brittle material, however, the ductility of reinforced concrete depends on the design. Reinforced concrete beams are designed to be under-reinforced to make them ductile, but this is not required for columns. Thus, in the structural design of reinforced concrete rigid frames, it is necessary to design the columns to be stronger than the beams so that plastic hinges can be formed in the beams. In this way, a reinforced concrete rigid frame is forced into ductile behaviour.

In the case of tall buildings, when designed for strength considerations only, the biggest disadvantage in rigid frame systems is the magnitude of lateral drift, which causes discomfort to occupants and damage to non-structural elements. There are two causes of lateral drift: the first is the deformation due to cantilever bending of the building (bending deformation), which is approximately 20 per cent of the total lateral drift (Figure 3.4a). The second is that of the deformation due to bending of the beams and columns (shear deformation), approximately 65 per cent is due to the bending of the beams, and 15 per cent to the columns, totalling approximately 80 per cent of the total lateral drift (Figure 3.4b) (Schueller, 1977).

Rigid frame systems efficiently and economically provide sufficient stiffness to resist wind and earthquake induced lateral loads in buildings of up to about 25 storeys. Some examples of tall buildings using the rigid frame system with steel structural material include:

- the 12-storey, 55 m high Home Insurance Building (Chicago,1885) (Figure 1.2) and
- the 21-storey, 94 m high Lever House (New York, 1952) (Figure 3.5)

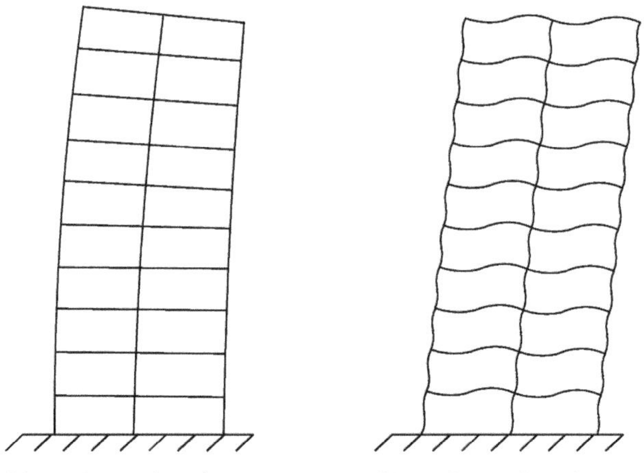

(a) Deformation due to cantilever bending

(b) Deformation due to the bending of the columns and beams

FIGURE 3.4 Lateral drift in rigid frame systems

FIGURE 3.5 Lever House, New York, USA, 1952

and with reinforced concrete structural material include:

- the 16-storey, 65 m high Ingalls Building (Cincinnati, 1903) (Figure 3.1)

3.3 Flat plate/slab systems

Flat plate/slab systems are used in reinforced concrete buildings. This system consists of beamless floor slabs of constant thickness and columns. Shear walls also can be placed in addition to or instead of the columns (Figure 3.6a). Column capitals (Figure 3.6b) or gussets (Figure 3.6c) can be placed on the upper ends of the columns in order to reduce the punching effect created by shear forces in the connections between the columns and slabs. Using a flat ceiling instead of one with beams, and

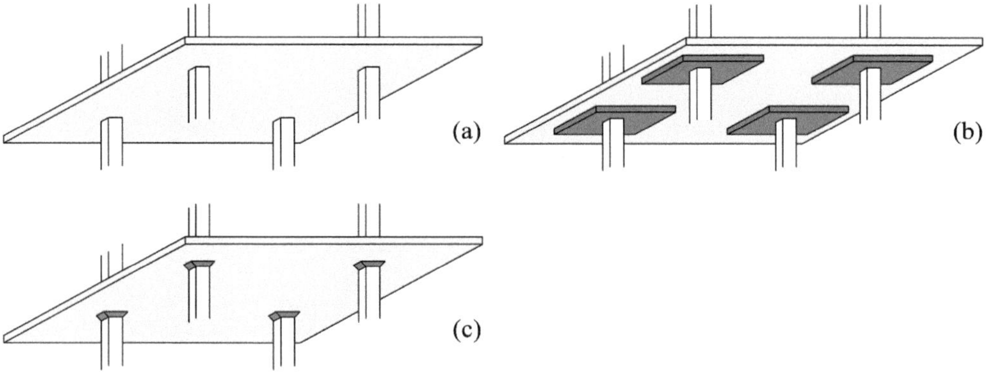

FIGURE 3.6 Flat plate/slab systems: (a) without column capitals, (b) with column capitals, (c) with gussets

thus attaining the maximum net floor height, is a major architectural advantage of this system.

In resisting lateral loads, flat plate/slab systems may be insufficient, compared with rigid frames. The reason for this is the shallow-wide-beam behaviour of the floor slab, with low bending/flexural rigidity. Thus real frame behaviour that has beams having sufficient depth cannot be achieved. The addition of shear walls to flat plate/slab systems mitigates this problem and increases the resistance against lateral loads.

Flat plate/slab systems efficiently and economically provide sufficient stiffness to resist wind and earthquake induced lateral loads in buildings of up to about 25 storeys.

3.4 Core systems

Core systems are used in reinforced concrete buildings. This system consists of a reinforced concrete core shear wall resisting all the vertical and lateral loads (Figure 3.7). In general, a core wall is an open core that is converted into a partially closed core by using floor beams and/or slabs so as to increase the lateral and torsional stiffness of the building. Although the behaviour of closed cores is ideal against building torsion under lateral loads, a partially closed core is used to approximate this for architectural reasons. Thus, a partially closed core is produced by supporting the open part of the core with beams and/or slabs having satisfactory strength against shear and bending.

In core systems, floor slabs are cantilevered from the core shear wall independently (Figure 3.8a), or else cantilevered modules of floor slabs are used (Figure 3.8b). In the case of cantilevered modules, floor slabs, except the bottom slab of each module, are cantilevered from the core shear wall and are supported by discontinuous perimeter columns down through the height of the modules. The bottom slab of each module is a strengthened cantilever floor slab which supports the perimeter columns of the upper storeys in the module.

FIGURE 3.7 Core system

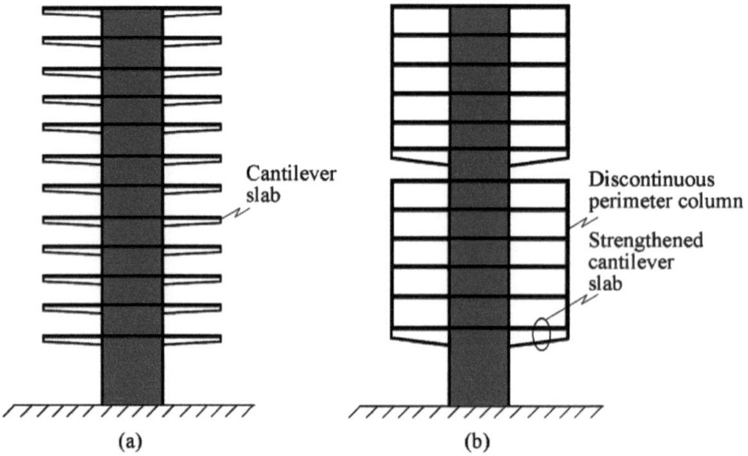

FIGURE 3.8 Slabs in core systems: (a) cantilever slabs, (b) strengthened cantilever slabs

The bending/flexural rigidity of the core in the core systems is limited by the flexural depth of the core. Thus in supertall buildings or in cases where the lateral load is very great, the bending/flexural rigidity of the building is not sufficient unless a mega core is used.

Core systems efficiently and economically provide sufficient stiffness to resist wind and earthquake induced lateral loads in buildings of up to about 20 storeys; however, "mega core systems" (Section 3.8), which are made with much thicker core shear walls than normal, can be used efficiently and economically in buildings of more than 40 storeys.

3.5 Shear wall systems

Shear wall systems are used in reinforced concrete buildings. This system consists of reinforced concrete shear walls, which can be perforated (with openings) or solid. Shear wall systems can be thought of as a vertical cantilever rigidly fixed at the base, and can resist all vertical and lateral loads on a building without columns (Figure 3.9). Owing to the nature of cantilever behaviour, the inter-storey drift between adjacent floors is greater in the upper floors than in the other floors. For this reason, in supertall buildings it is difficult to control the lateral drift at the building top.

Shear wall systems efficiently and economically provide sufficient stiffness to resist wind and earthquake induced lateral loads in buildings of up to about 35 storeys.

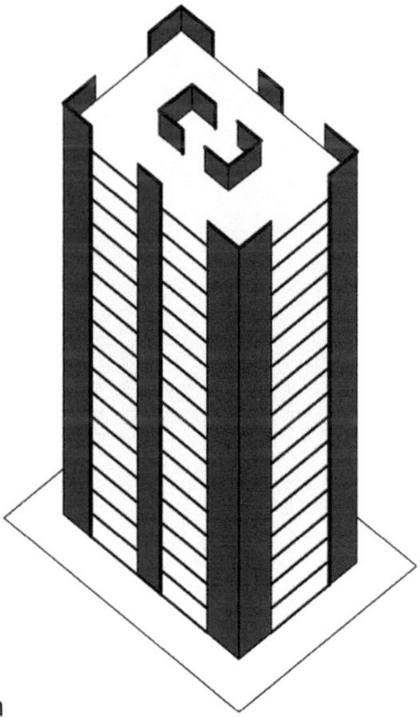

FIGURE 3.9 Shear wall system

3.6 Shear-frame systems

Rigid frame systems economically do not have sufficient resistance against lateral loads in buildings over 25 storeys because of bending on columns that causes large deformations. In this case, the total stiffness and so the economical height of the building can be increased by adding vertical shear trusses (braces) and/or shear walls to the rigid frame to carry the external shear induced by lateral loads (Figure 3.10).

This interactive system of frames and shear trusses and/or shear walls is called the "shear-frame system", and is quite effective against lateral loads (Figure 3.11). In this context, shear-frame systems can be divided into two types:

- shear trussed frame (braced frame) system (Figure 3.11a)
- shear walled frame system (Figure 3.11b).

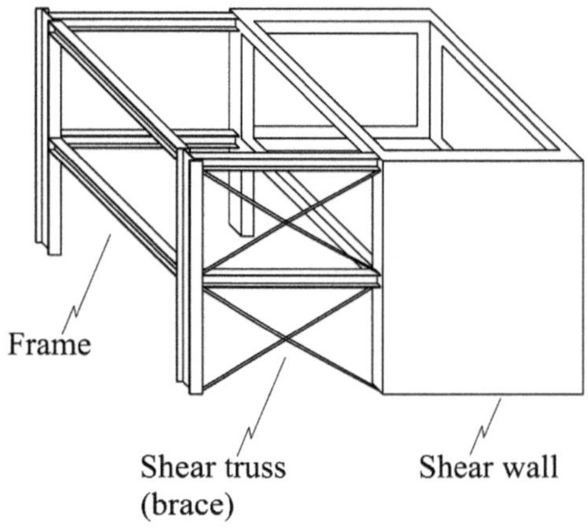

FIGURE 3.10 Rigid frame, shear truss (brace), and shear wall

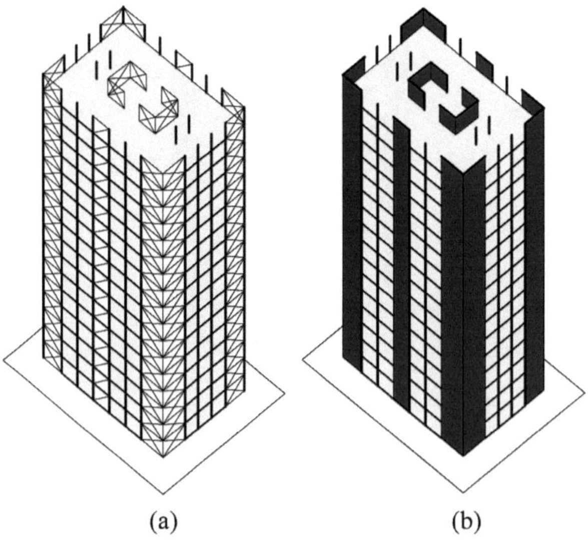

FIGURE 3.11 (a) Shear trussed frame (braced frame) system, (b) shear walled frame system

The shear trussed frame (braced frame) system consists of rigid frames and diagonal braces in the form of vertical trusses, while the shear walled frame system consists of rigid frames and reinforced concrete shear walls that are perforated or solid (Figure 3.13a).

It is also possible to design shear trusses (braces) and/or shear walls as cores that surround elevator shafts and stairwells (Figure 3.13b). In this case, shear-frame systems having structural cores can also be called as "core-frame systems", and likewise, shear trussed frame and shear walled frame systems having structural cores "core-trussed frame" and "core-walled frame" systems respectively. Closed and partially closed cores increase the stiffness of the building laterally and torsionally.

Rigid frames resist lateral loads by dissipating energy through their ductility and may undergo excessive lateral deformations. Under lateral loads, the slope of the deformed shape, in other words, the inter-storey drift between adjacent floors is higher in the lower storeys. Shear trusses and walls, despite being less ductile than the frame, dissipate energy while staying within elastic limits because of the larger size of the shear area subjected to the shear force caused by lateral loads, and exhibit smaller lateral deformations. Thus, ductility is not as important as it is with rigid frames. The slope of the deformed shape, in other words, the inter-storey drift between adjacent floors, is higher in the upper storeys, and is greatest at the top. The disadvantages of the frame compared with the shear truss or wall, and of the shear truss or wall compared with the frame, are compensated by one another in a system where they are used together, in which the frame contributes to the shear truss or wall in the upper storeys, while the shear truss or wall contributes to the frame in the lower storeys. In this way, the shear-frame system exhibits very effective behaviour against lateral loads by giving the structure a greater stiffness than a system of "shear truss / shear wall" or "rigid frame" acting alone (Figure 3.12).

When shear trusses and shear walls are designed as cores that surround elevator shafts and stairwells, they form partially closed cores since the cross-section of the core is not completely but partially closed by beams and/or floor slabs (Figure 3.13b). The partially closed cores in general are arranged in rectangular or circular shapes. The effort is made to approximate the behaviour of a closed core by strengthening beams and/or floor slabs in the open part of the core, providing sufficient stiffness against shear and bending.

The location and shape of the shear trusses and shear walls affect their performance under lateral loads to an important degree. By arranging them in such a way that the resultant lateral force acts through the centre of rigidity of the building, shear-frame

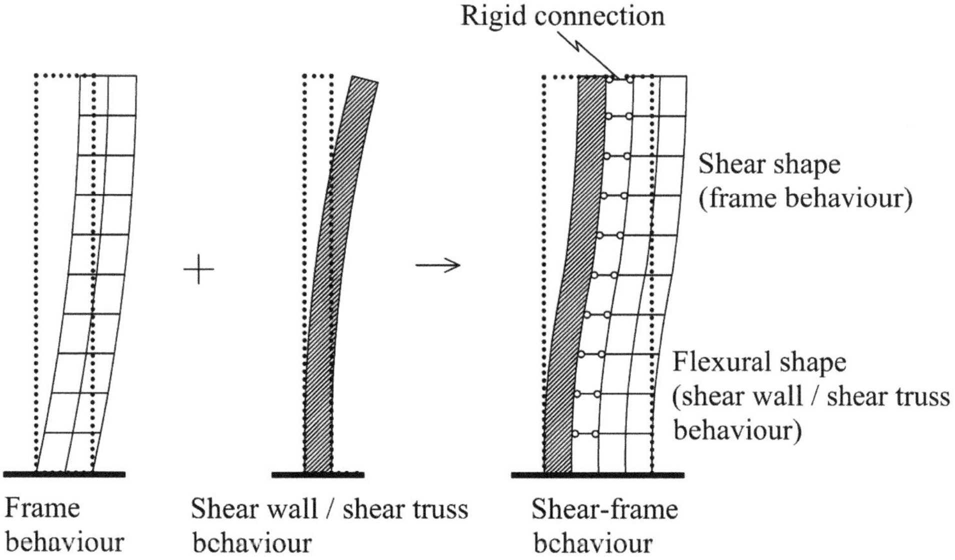

FIGURE 3.12 The behaviour of the shear-frame system under lateral loads

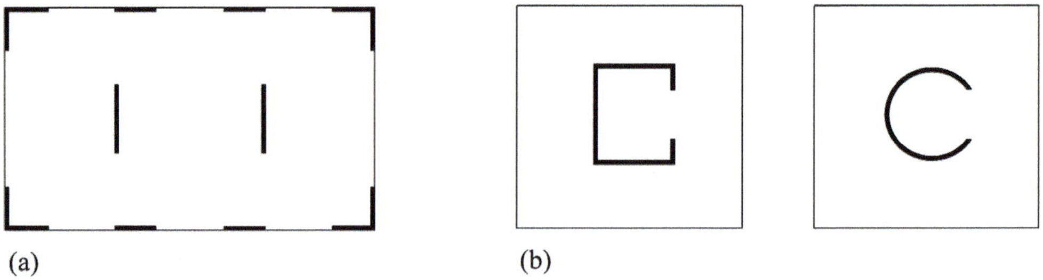

FIGURE 3.13 (a) Shear trusses / shear walls in plan, (b) partially closed cores in plan

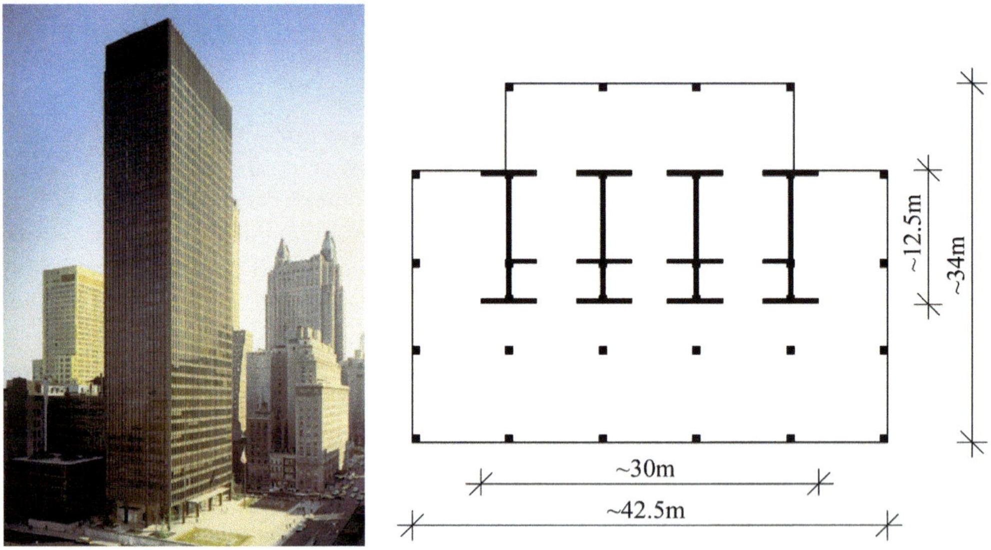

FIGURE 3.14 Seagram Building, New York, USA, 1958
(photo courtesy of Antony Wood / CTBUH)

systems are not subject to torsion. Otherwise, torsion occurs, and the torsional forces must also be taken into account. The most effective behaviour against torsion in the shear trusses and shear walls is ensured by partially closed cores.

Shear-frame systems efficiently and economically provide sufficient stiffness to resist wind and earthquake induced lateral loads in buildings of more than (as well as below) 40 storeys. In shear-frame systems, shear trusses and shear walls may be used together with rigid frame, as in the 38-storey, 157 m high Seagram Building (New York, 1958) (Figure 3.14), with its composite structural system, designed by Mies van der Rohe. In the 38-storey Seagram Building, up to the seventeenth floor, reinforced concrete shear walls, and in the upper storeys steel shear trusses were used (Ali and Moon, 2007).

3.6.1 Shear trussed frame (braced frame) systems

Shear trussed frame (braced frame) systems consist of rigid frames and braces in the form of vertical trusses (Figure 3.11a). Diagonal brace elements between the columns of the rigid frame create a truss frame at that bay where those columns act as vertical continuous chords. Columns, beams and braces are generally made of steel, sometimes composite, but rarely are of reinforced concrete. Because the lateral loads can act in all directions and/or cyclic, brace elements are exposed to both tension and compression. Reinforced concrete braces are weak in tension due to the inherent characteristic of concrete. Therefore in general, steel but sometimes composite braces are preferred. On the other hand, buckling is a critical factor for steel braces and they must be sized accordingly.

The diagonal brace elements can be single or double. Brace elements (diagonals) are designed taking buckling into account, or else double diagonal braces (X-braces, chevron-braces, knee-braces) are used and designed in such a way that, according to the direction of the lateral force, when tension occurs in one of the brace elements it is assumed that the other brace element buckles and compression force is not developed in the brace elements. Likewise, in cases where braces are of reinforced concrete, their inability to take tensile forces is taken into account, and double diagonal braces are used assuming that tension force is not developed in the brace elements, namely, ignored, by designing brace diagonals as compression members. In other words, only half the braces carry lateral loads at a time.

Architecturally, shear truss bracing can be divided into four groups (Figure 3.15a):

- diagonal-bracing
- x-bracing (cross-bracing)
- chevron-bracing (v-bracing)
- knee-bracing.

Structurally, shear truss bracing can be divided into two groups (Figure 3.15b):

- concentric-bracing
- eccentric-bracing.

Compared with "Knee" and "Chevron" braces, the "X" and "diagonal" braces are an architectural obstacle, reducing the field of view through their openings and making it difficult to install doors and windows. Because of this, "X" and "diagonal" braces are preferred in locations where openings are not necessary, such as partition walls, elevator shafts and stairwells.

Structurally, braced frames can be divided into two groups as either concentric-braced frames or eccentric-braced frames by the type of bracing. In concentric-braced frames, beams, columns and braces meet at a common connection so that the member forces are primarily axial. In eccentric-braced frames, braces are placed eccentrically to the beam-column connection (diagonals are placed from beam-column to beam or beam to beam) which creates bending moment and so flexure In the beam. Structurally, concentric-braced frames contribute to lateral stiffness within

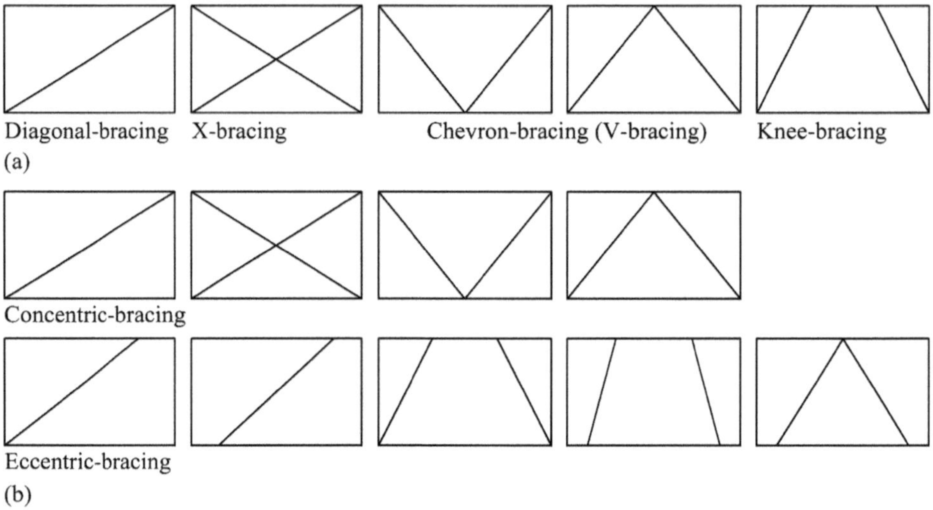

FIGURE 3.15 Types of bracing

elastic limits, while eccentric-braced frames do so within both elastic and inelastic limits. Although the lateral stiffness and so deformation of eccentric-braced frame systems are not as great as in concentric-braced frame systems, they are preferred in seismic regions because of their energy dissipation capacity and ductility. In resisting ultimate loads, energy produced by the external shear is dissipated/absorbed by ensuring ductility through bending and shear in the lower and upper beams of shear truss (truss frame).

The 21-storey, 92 m high Masonic Temple (Chicago, 1892), designed by the architects Burnham and Root in Chicago in 1892, was the first tall building in which a shear trussed frame system was used.

The shear trussed frame system has been used in many tall buildings which held the title of "the world's tallest building" in its time, including the 77-storey, 319 m high Chrysler Building (New York, 1930) (Figure 3.16), and the 102-storey, 381 m high Empire State Building (New York, 1931) (Figure 3.17).

3.6.2 Shear walled frame systems

Shear walled frame systems consist of rigid frames and reinforced concrete shear walls that are perforated or solid (Figure 3.11b). In general, shear walls are of reinforced concrete; occasionally of composite formed by concrete encased structural steel, or of steel plates. Columns and beams are reinforced concrete, steel or composite.

Some examples of tall buildings using the shear walled frame system with reinforced concrete structural material include:

- the 32-storey, 127 m high Pirelli Building (Milan, 1958) (the first reinforced concrete building utilising the interactive system of rigid frames and shear walls)

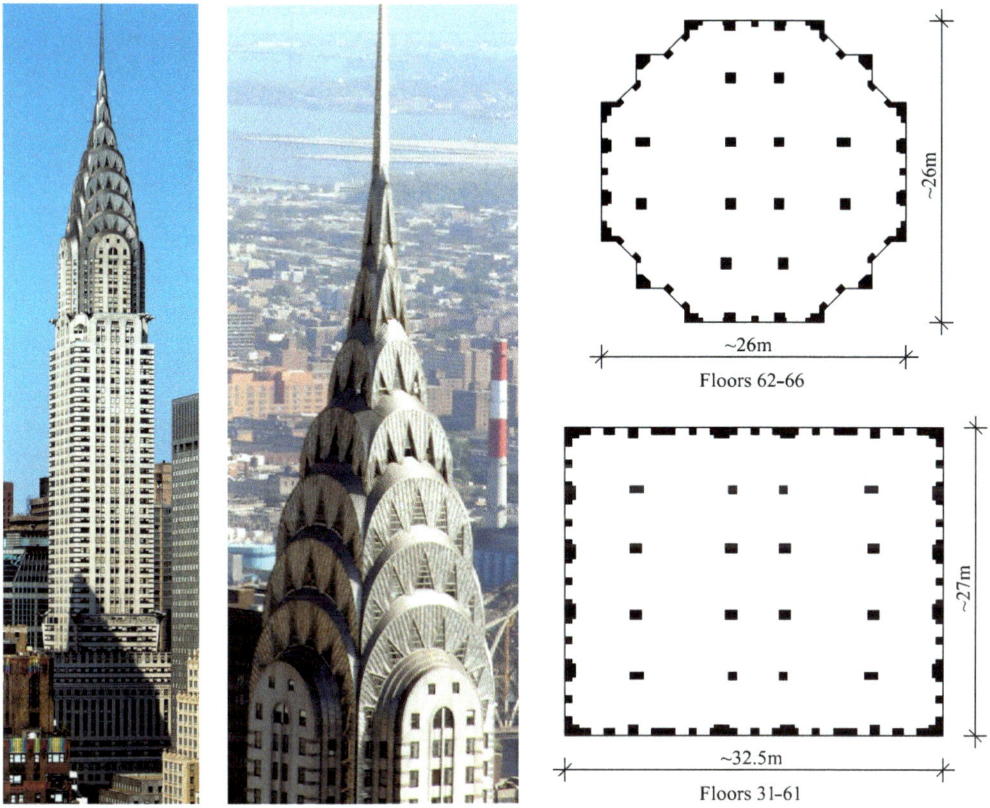

FIGURE 3.16 Chrysler Building, New York, USA, 1930
(photo on right courtesy of Antony Wood/CTBUH)

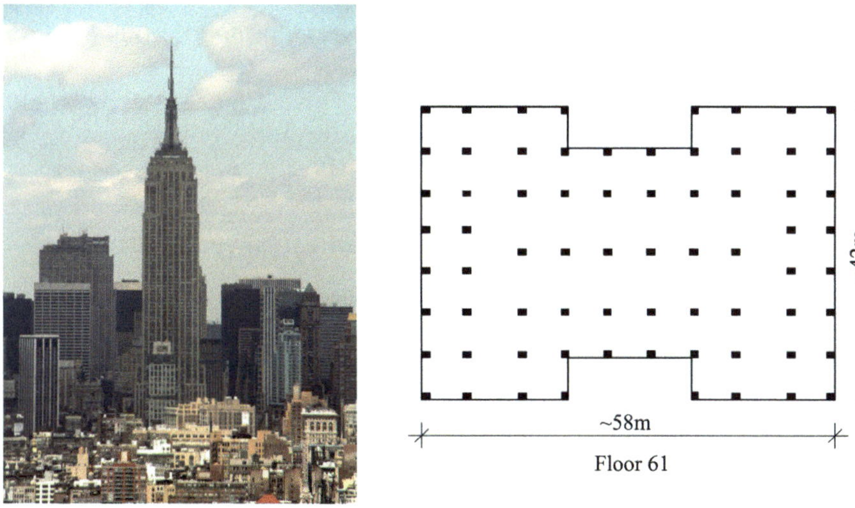

FIGURE 3.17 Empire State Building, New York, USA, 1931
(photo courtesy of Antony Wood/CTBUH)

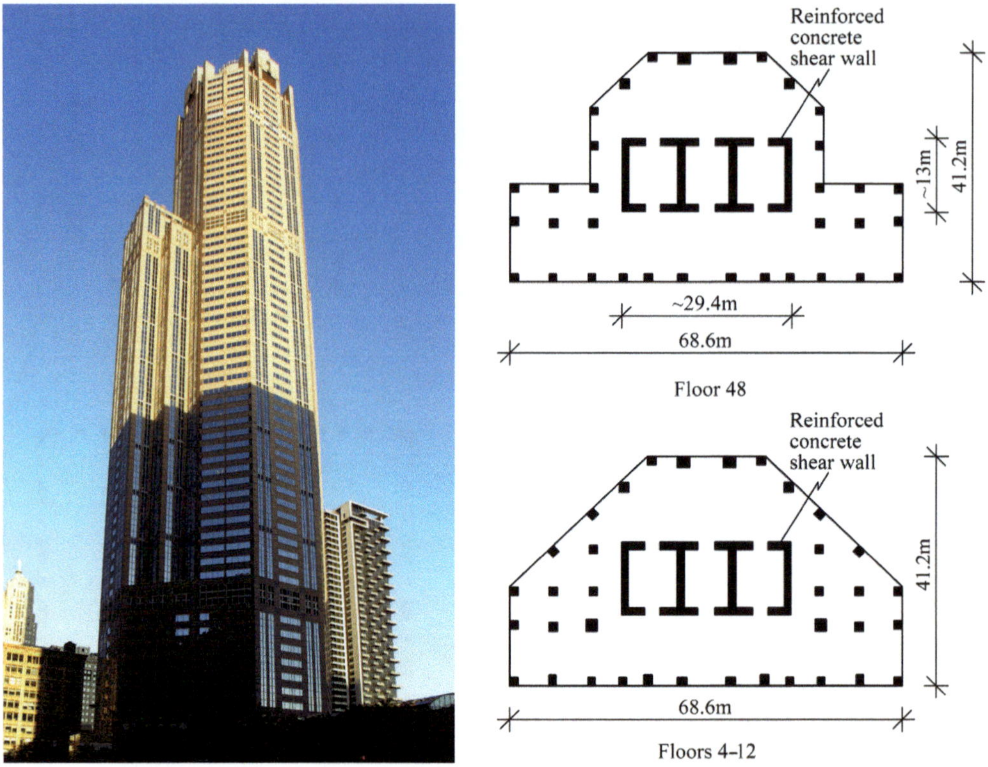

FIGURE 3.18 311 South Wacker Drive, Chicago, USA, 1990
(photo courtesy of Marshall Gerometta/CTBUH)

- the 35-storey, 145 m high Cook County Administration Building (formerly Brunswick Building) (Chicago, 1964)
- the 65-storey, 293 m high 311 South Wacker Drive (Chicago, 1990) (Figure 3.18)
- the 30-storey, 267 m high Al Faisaliah Center (Riyadh, 2000) (Figure 3.19)
- the 43-storey, 148 m high Strata (London, 2010) (Figure 3.20).

3.7 Mega column (mega frame, space truss) systems

Mega column systems consist of reinforced concrete or composite columns and/or shear walls with much larger cross-sections than normal, running continuously throughout the height of the building. In this system, mega columns and/or mega shear walls can resist all the vertical and lateral loads (Figure 3.21).

In mega column systems, horizontal connections are of primary importance. Due to the probable insufficiency of floor slabs acting as rigid floor diaphragms, to support this behaviour of restraining the columns laterally, belts, vierendeel frames, and mega braces are used. In this way, all external mega columns and/or shear walls are connected together to participate in the lateral stiffness of the structure (Figure 3.21a).

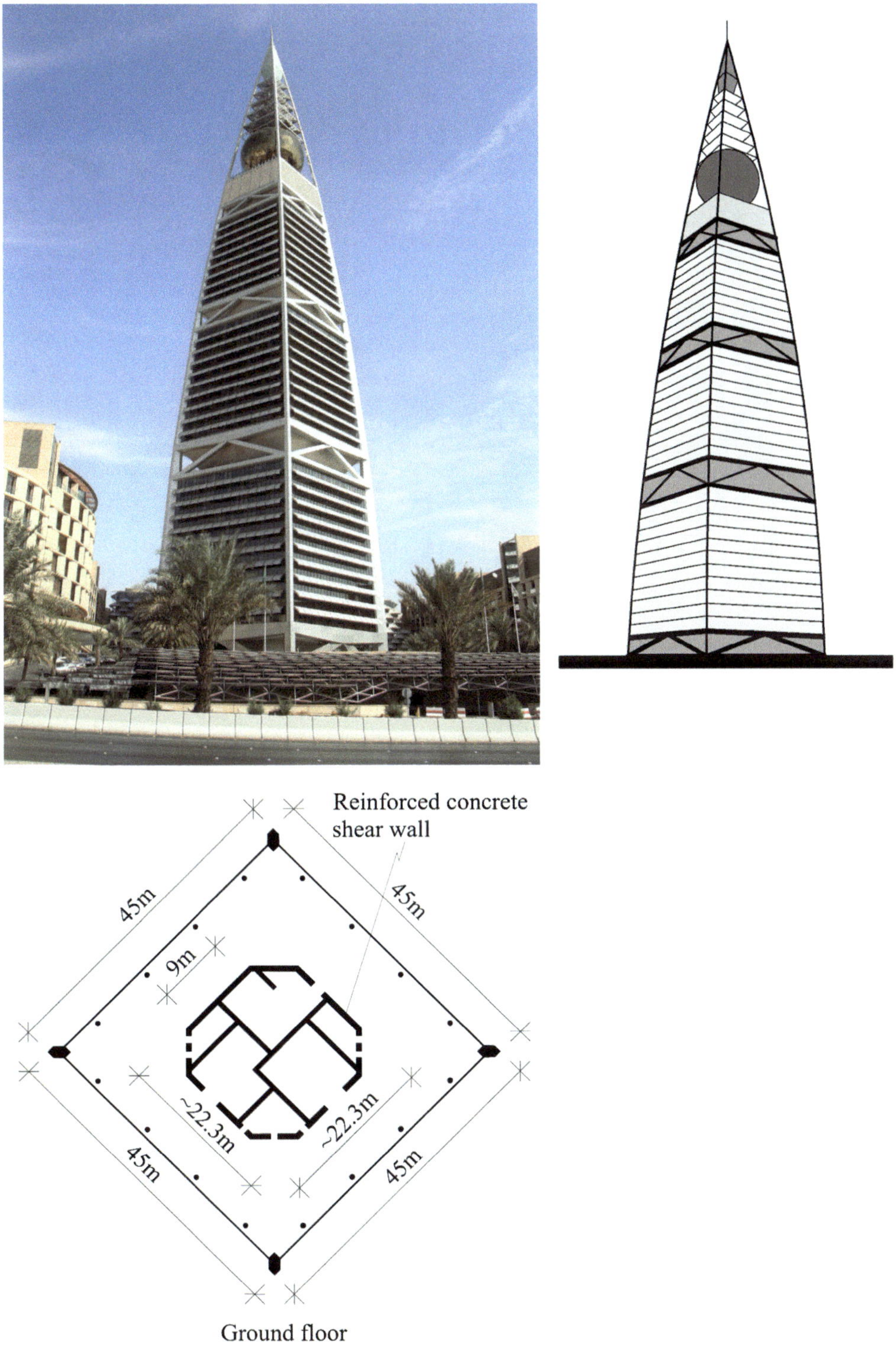

FIGURE 3.19 Al Faisaliah Center, Riyadh, Saudi Arabia, 2000
(photo courtesy of Adrian Peret, adrian.peret@gmail.com)

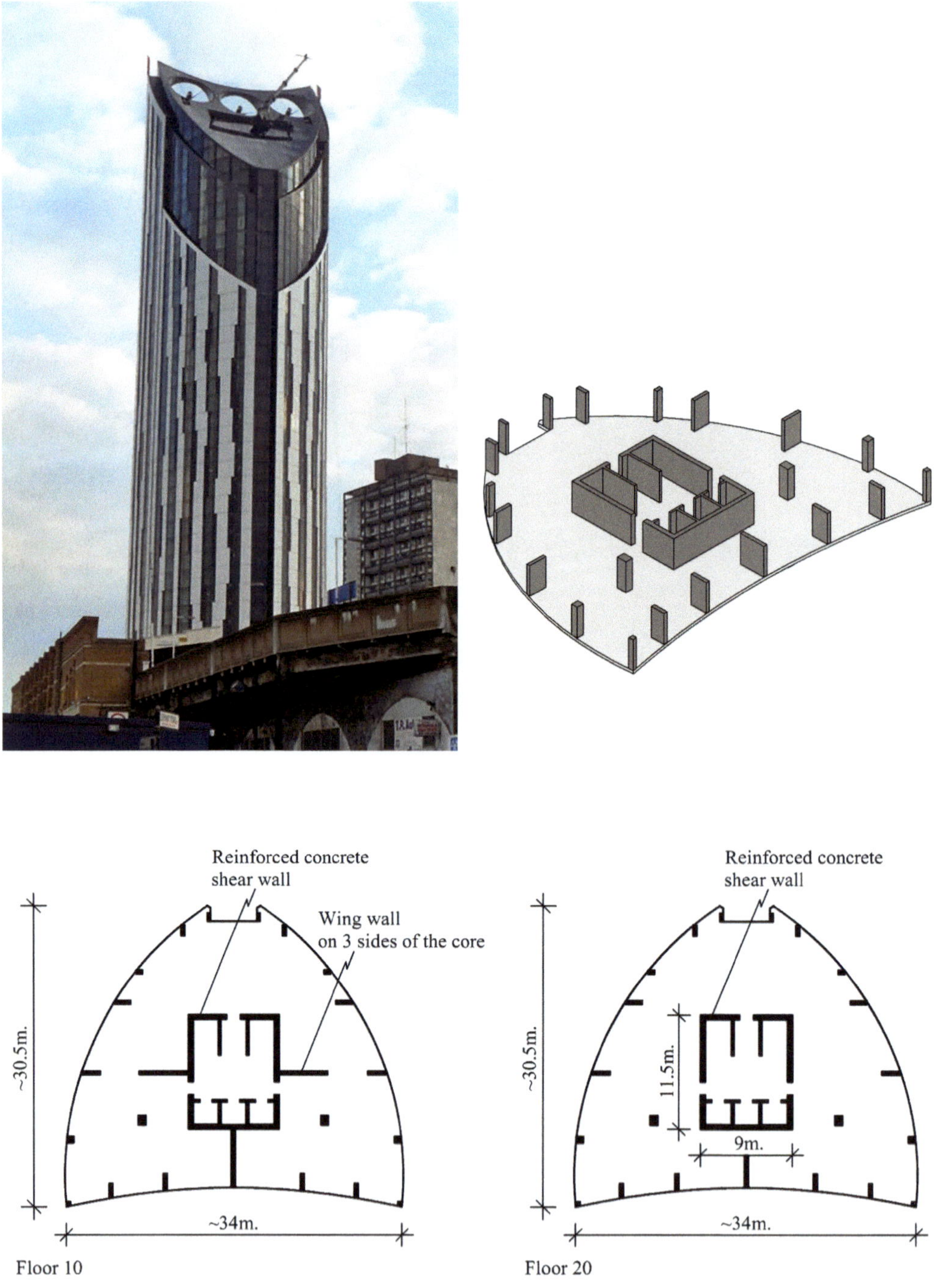

FIGURE 3.20 Strata, London, UK, 2010

Belts and vierendeel frames consist of at least one storey depth horizontal shear trusses or shear walls that located at least two or more levels throughout the height of the building as in the case of the Commerzbank Tower (Frankfurt, 1997) (Figure 3.23), which has 6 mega shear walls connected with vierendeel frames. Mega braces are multi-storey diagonals that are placed continuously throughout the height of the building as in the case of the Bank of China Tower (Hong Kong, 1990) (Figure 3.71), which has 4 composite mega columns connected with both mega braces and belts.

According to the authors, mega column systems, in their function and appearance, can also be named as "mega frame systems" (Figure 3.21a); likewise, in some cases where there are mega braces supporting the mega columns, being reminiscent of a three dimensional truss, they can also be named as "space truss systems" as in the case of the Bank of China Tower (Hong Kong, 1990) (Figure 3.21b).

Mega column systems efficiently and economically provide sufficient stiffness to resist wind and earthquake induced lateral loads in buildings of more than 40 storeys. Some examples of tall buildings using the mega column system with composite structural material include:

- the 73-storey, 346 m high Center (Hong Kong, 1998) (Figure 3.22), which has 12 composite mega columns, of which the largest have square cross-sections of 2.5×2.5 m at the ground floor

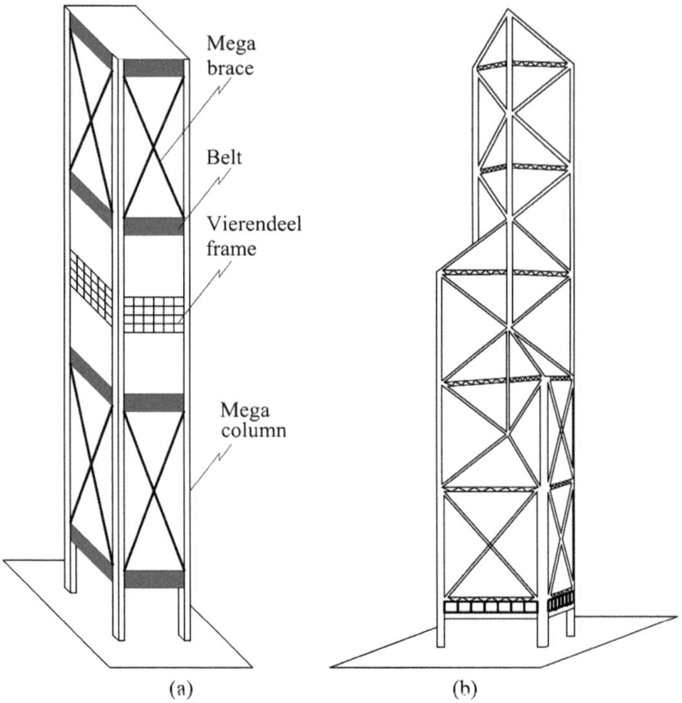

FIGURE 3.21 Mega column (mega frame, space truss) system

- the 56-storey, 259 m high Commerzbank Tower (Frankfurt, 1997) (Figure 3.23), which has 6 mega composite shear walls having cross-sections of approximately 1.2×7.5 m
- the 72-storey, 367 m high Bank of China Tower (Hong Kong, 1990) (Figure 3.71), which has 4 composite mega columns with hexagonal cross-sections at the ground floor that are approximately 3.50 m on their two longest sides. Considering the function and appearance, the structural system of the Bank of China Tower is categorized as space truss system by Ali and Armstrong (1995) and Ali and Moon (2007).

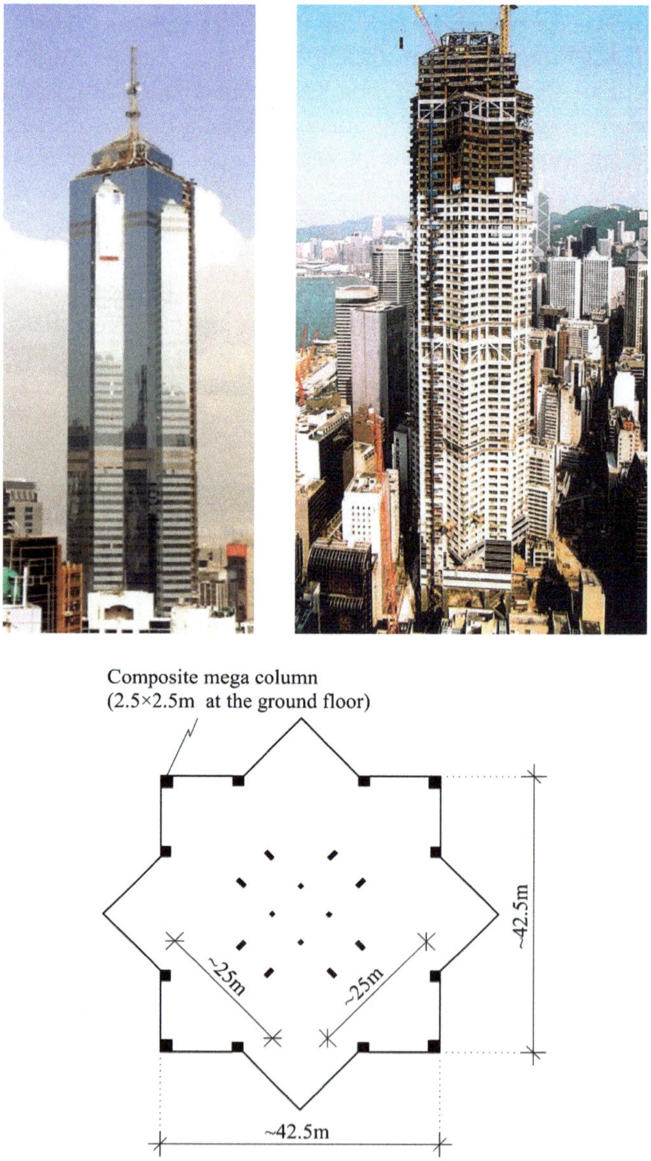

FIGURE 3.22 The Center, Hong Kong, China, 1998
(photos courtesy of Derek Forbes)

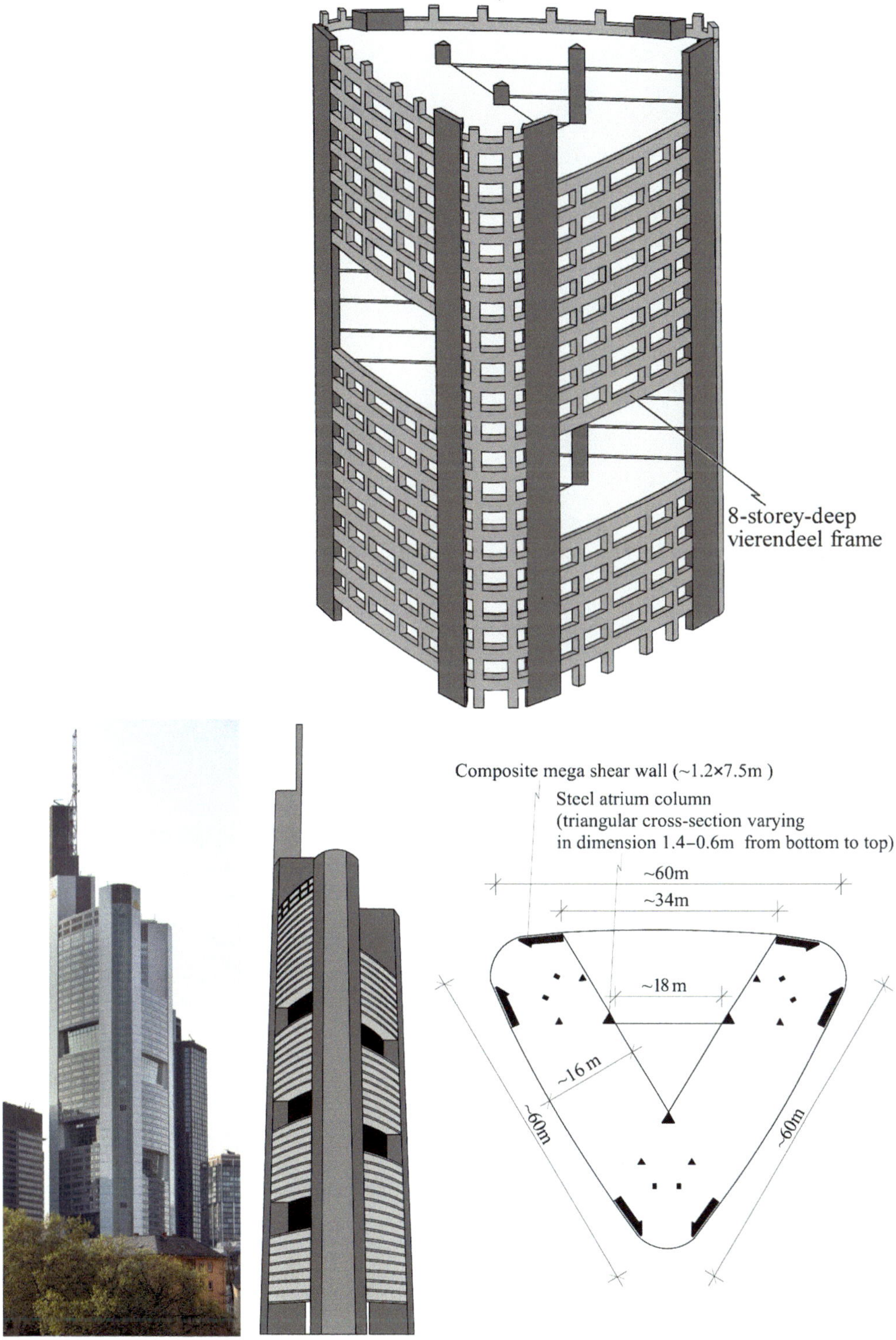

FIGURE 3.23 Commerzbank Tower, Frankfurt, Germany, 1997

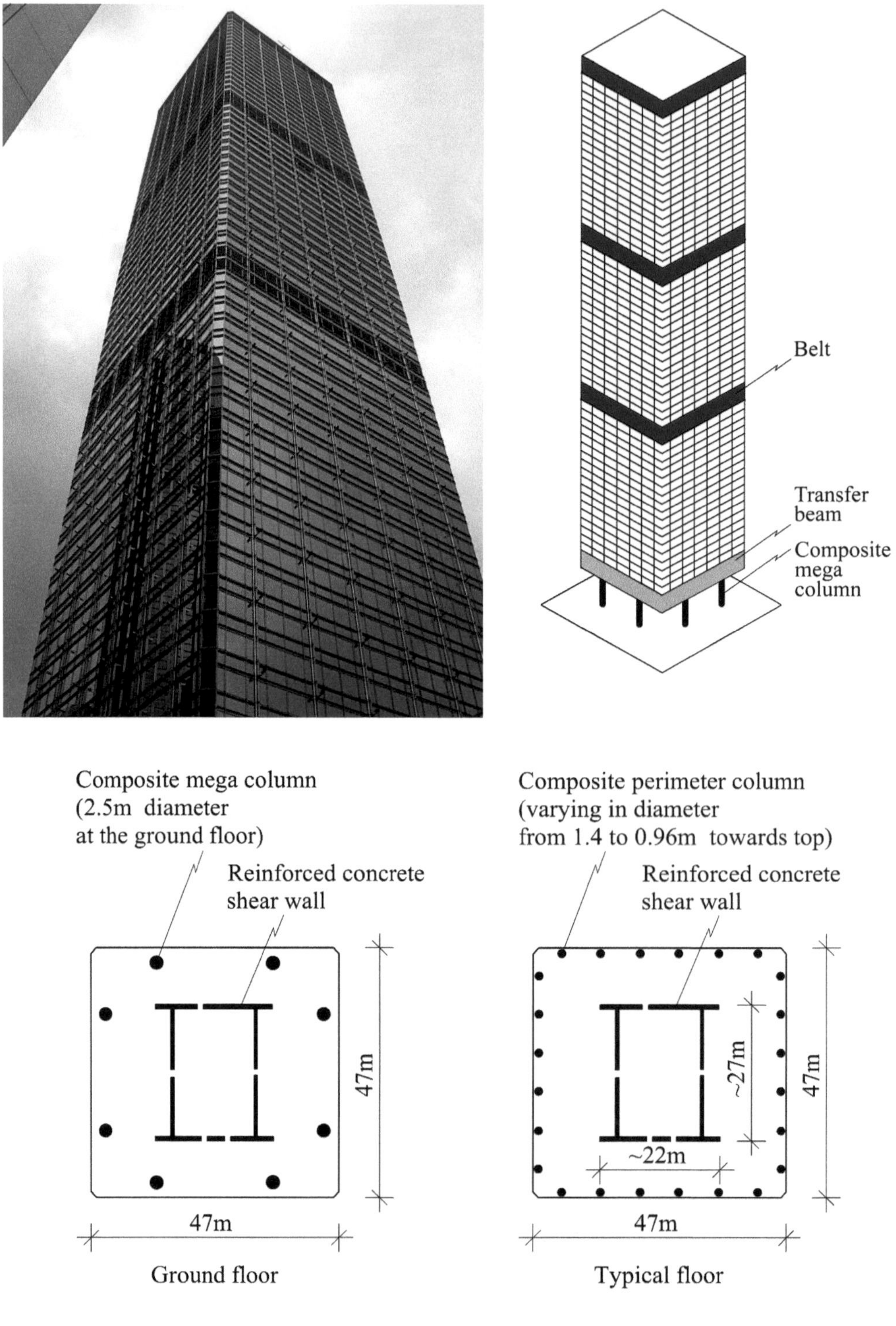

FIGURE 3.24 Cheung Kong Centre, Hong Kong, China, 1999
(photo courtesy of Niels Jakob Darger)

Mega columns can also be used solely to provide large spaces at the building entrance, as an aid to the main structural system for the levels above the entrance, without running continuously throughout the height of the building. As the number of mega columns at the entrance is much lower than the number of columns on the upper storeys, the structural transition between them is achieved using deep transfer beams. In such cases, the cross-sectional dimensions of the column at the entrance are large enough for it to be classified as a "mega column", but the structural system cannot be classified as a "mega column system".

Tall buildings where this approach has been used include the 63-storey, 283 m high Cheung Kong Centre (Hong Kong, 1999) (Figure 3.24), which has an outriggered frame system and 8 composite mega columns at the ground floor with 2.5 m diameter circular cross-sections, and the 59-storey, 279 m high Citigroup Center (New York, 1977) (Figure 3.66), which has a trussed-tube system and 4 steel mega columns at the ground floor with rectangular cross-sections of approximately 6.5×7 m.

Mega columns, in cases where they run continuously throughout the height of the building, can be used with an outriggered frame system or a tube system. In such cases, when they are used for a purpose such as reducing the number of columns, the structural system cannot be classified as "mega column system", since the mega columns are not the only structural elements that resist the external loads. Tall buildings with outriggered frame systems include:

- the 101-storey, 508 m high Taipei 101 (Taipei, 2004) (Figure 3.36), which has 8 composite mega columns at the ground floor with rectangular cross-sections of 2.4×3 m
- the 88-storey, 421 m high Jin Mao Building (Shanghai, 1999) (Figure 3.40), which has 8 composite mega columns at the ground floor with rectangular cross-sections of 1.5×4.9 m
- the 88-storey, 412 m high Two International Finance Centre (Hong Kong, 2003) (Figure 3.41), which has 8 composite mega columns at the ground floor with rectangular cross-sections of 2.5×3.5 m.

3.8 Mega core systems

Mega core systems consist of reinforced concrete or composite core shear walls with much larger cross-sections than normal, running continuously throughout the height of the building (Figure 3.25). Since the mega core can resist all vertical and lateral loads in this system, there is no need for columns or shear walls on the perimeter of the building. In mega core systems, floor slabs are cantilevered from the core shear wall (Figure 3.25a). Mega core systems can also be used with strengthened cantilever slabs (Figure 3.25b). In this case, floor slabs are supported by the core shear walls and discontinuous perimeter columns. Perimeter columns are supported by strengthened cantilever slabs repeated on some storeys. Strengthened cantilever slabs protrude from the core, and are strengthened in order to support the load coming from the storeys above.

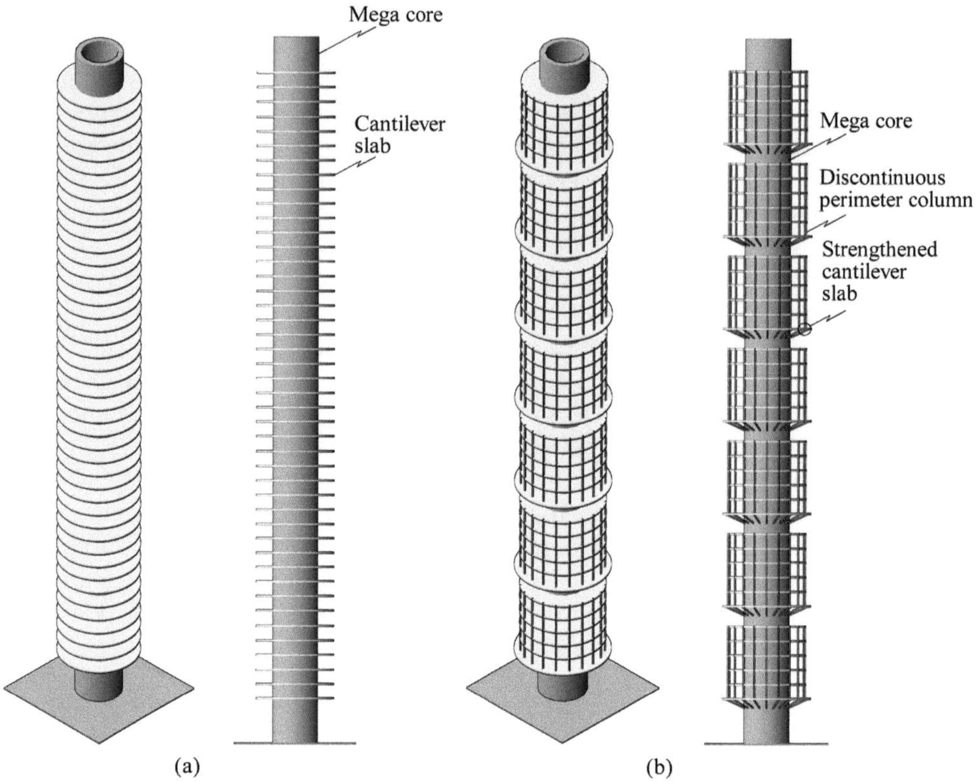

FIGURE 3.25 Slabs in the mega core system: (a) cantilever slab, (b) supported cantilever slab

Mega core systems efficiently and economically provide sufficient stiffness to resist wind and earthquake induced lateral loads in buildings of more than 40 storeys. Some examples of tall buildings using the mega core system with reinforced concrete structural material include:

- the 36-storey, 300 m high Aspire Tower (Doha, 2006) (Figure 3.26) which has a reinforced concrete core shear wall having circular cross-section with an external diameter varying between 18 to13 m (from bottom to top) and thickness varying between 2 to 1 m (from bottom to top)
- the 52-storey, 235 m high 8 Shenton Way (Singapore, 1986) (Figure 3.27), which has a reinforced concrete core shear wall having circular cross-section with an external diameter of 25 m and thickness varying between 1.65 to 1 m (from bottom to top)
- the 57-storey, 190 m high HSB Turning Torso (Malmö, 2005) (Figure 3.28) which has a reinforced concrete core shear wall having circular cross-section with an external diameter varying between 15.6 to 11.4 m (from bottom to top) and thickness varying between 2.5 to 0.4 m (from bottom to top).

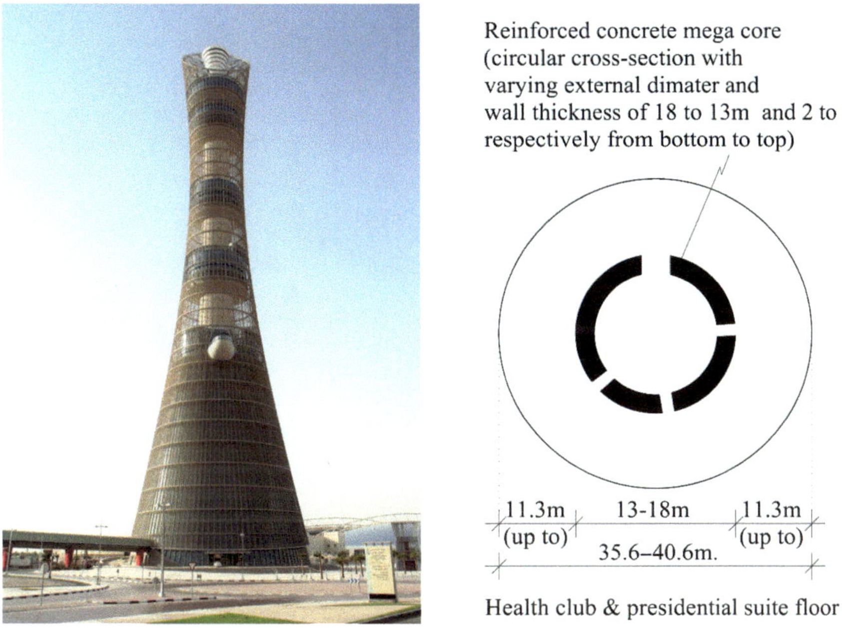

FIGURE 3.26 Aspire Tower, Doha, Qatar, 2006
(credit for Photo: CTBUH)

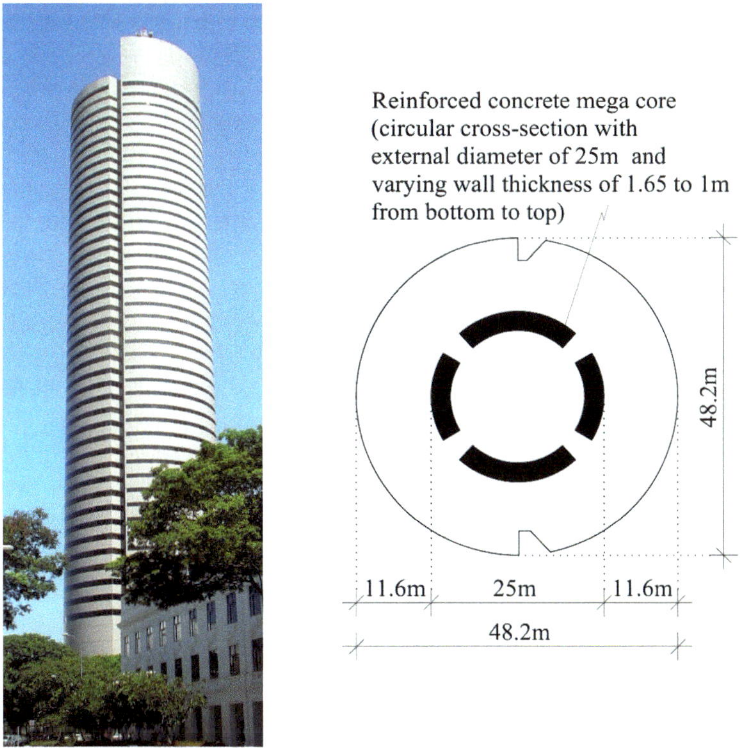

FIGURE 3.27 8 Shenton Way, Singapore, Singapore, 1986

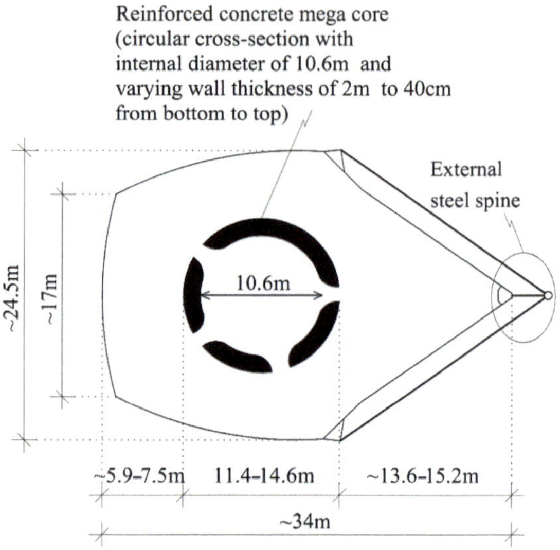

FIGURE 3.28 HSB Turning Torso, Malmö, Sweden, 2005
(photo courtesy of Santiago Calatrava/Samark Architecture & Design)

3.9 Outriggered frame systems

Outriggered frame systems have been developed by adding outriggers to shear-frame systems with core (core-frame systems) so as to couple the core with the perimeter (exterior) columns. The outriggers are structural elements connecting the core to the perimeter columns at one or more levels throughout the height of the building so as to stiffen the structure (Figure 3.29). An outrigger consists of a horizontal shear truss or shear wall (or deep beam). This structural element is a horizontal extension of the core shear truss/wall to the perimeter columns in the form of a knee. To make them sufficiently effective, outriggers are at least one storey deep, and have a high flexural and shear rigidity (adequately stiff in flexure and shear). Because the outriggers affect the interior space, they are generally located at the mechanical equipment floors in order not to hinder the use of normal floors.

The outriggers, which are connected rigidly to the core and by hinges to the perimeter columns, increase the effective flexural depth and so the flexural stiffness of the system in the direction of bending under lateral loads by enabling the core to receive support from the perimeter columns. The outrigger supports the core shear truss/wall against bending, creating axial tension and compression on the perimeter columns. In this way, the cantilever tube behaviour of the system is ensured, and the stiffness of the shear-frame system is increased, while reducing the lateral drift of the building to a significant degree.

At the levels of the outriggers, connecting the perimeter columns to each other with belts, improves the efficiency of the system by equalising the axial column loads along the perimeter. In this manner, the column, which is connected to the core by the

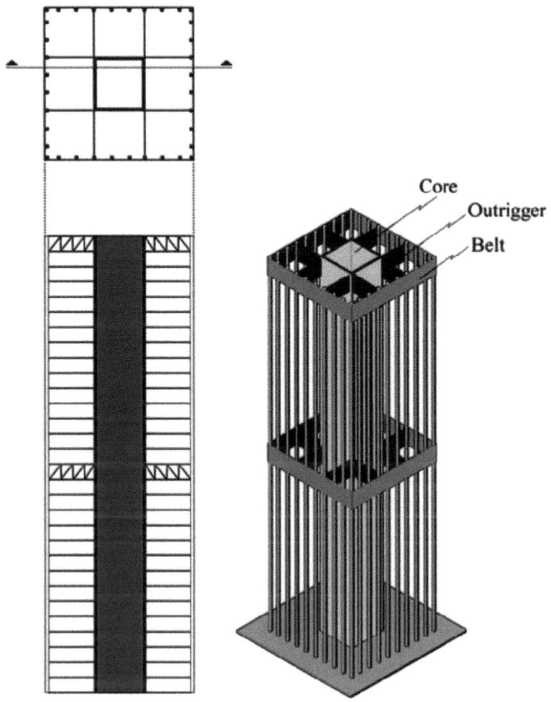

FIGURE 3.29 Outriggered frame system

outrigger, distributes the axial load effect of the outrigger to other columns by means of the belt. A belt consists of a horizontal shear truss or shear wall (or deep beam) adequately stiff in flexure and shear, and of equal depth to the outrigger (Figure 3.29). In this way, all perimeter columns are connected together to participate in supporting the outriggers. Belts are used not only in the abovementioned conventional outrigger systems, but also used in the "virtual" outrigger systems. Virtual outrigger concept takes advantage of floor diaphragms to eliminate direct connection of core and perimeter columns by outriggers. A virtual outrigger consists of belt, and floor slabs engaged by belt. In this manner, the problem associated with the space occupied by the conventional outriggers is avoided. Efficiency of the virtual outriggers depends on the rigidity of the belt and floor slabs at belt levels.

In cases where an outrigger is used at a single level throughout the height of the building, the most effective, and for this reason the optimum location for the outrigger is approximately 40–60 per cent of the building height (Smith and Coull, 1991; Taranath, 1998) (Section 3.9.2).

There is a relation between the number of levels where outriggers are used throughout the height of the building and their optimum locations. The optimum location of "n" number of outriggers used at levels throughout the height of the building can be given approximately by the formula 1/(n+1), 2/(n+1) ... n/(n+1) (Smith and Coull, 1991) (Section 3.9.2). The optimum locations for outriggers at one or two levels throughout the height of the building using various assumptions are calculated in Sections 3.9.1 and 3.9.2.

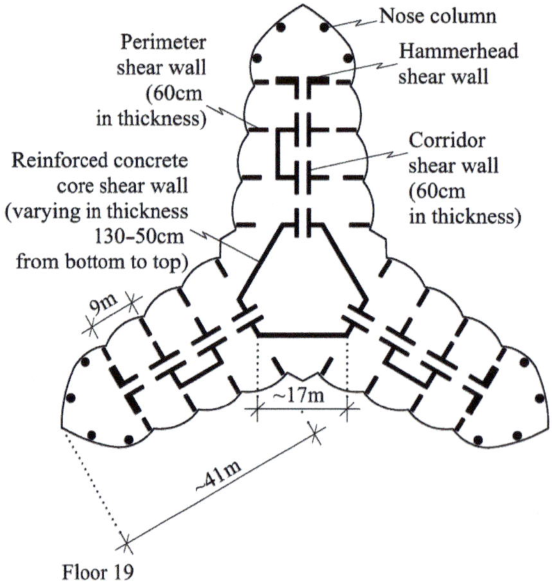

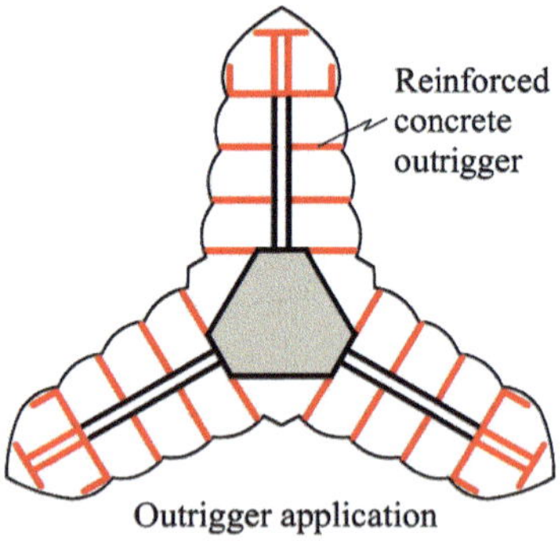

FIGURE 3.30 Burj Khalifa, Dubai, U.A.E, 2010
(photo courtesy of Adrian Peret, adrian.peret@gmail.com)

Addition of each new outrigger level increases the stiffness of the building, but by a smaller amount than the increase at the preceding level (Smith and Coull, 1991) (Section 3.9.2).

Outriggered frame systems efficiently and economically provide sufficient stiffness to resist wind and earthquake induced lateral loads in buildings of more than 40 storeys. Some examples of tall buildings using the outriggered frame system with reinforced concrete structural material include:

- the 163-storey, 828 m high Burj Khalifa (Dubai, 2010) (Figure 3.30)
- the 88-storey, 452 m high Petronas Twin Towers (Kuala Lumpur, 1998) (Figure 3.31)
- the 98-storey, 423 m high Trump International Hotel & Tower (Chicago, 2009) (Figure 3.32)
- the 91-storey, 297 m high Eureka Tower (Melbourne, 2006) (Figure 3.33)
- the 66-storey, 288 m high Plaza 66 (Shanghai, 2001) (Figure 3.34)

and with composite structural material include:

- the 121-storey, 632 m high Shanghai Tower (Shanghai, under construction) (Figure 3.35)
- the 101-storey, 508 m high Taipei 101 (Taipei, 2004) (Figure 3.36)
- the 101-storey, 492 m high Shanghai World Financial Center (Shanghai, 2008) (Figure 3.37)
- the 108-storey, 484 m high International Commerce Centre (ICC) (Hong Kong, 2010) (Figure 3.38)
- the 66-storey, 450 m high Zifeng Tower (Nanjing, 2010) (Figure 3.39)
- the 88-storey, 421 m high Jin Mao Building (Shanghai, 1999) (Figure 3.40)
- the 88-storey, 412 m high Two International Finance Centre (Hong Kong, 2003) (Figure 3.41)
- the 69-storey, 384 m high Shun Hing Square (Shenzhen, 1996) (Figure 3.42)
- the 52-storey, 319 m high New York Times Tower (New York, 2007) (Figure 3.43)
- the 63-storey, 283 m high Cheung Kong Centre (Hong Kong, 1999) (Figure 3.24)
- the 73-storey, 230 m high World Tower (Sydney, 2004) (Figure 3.44).

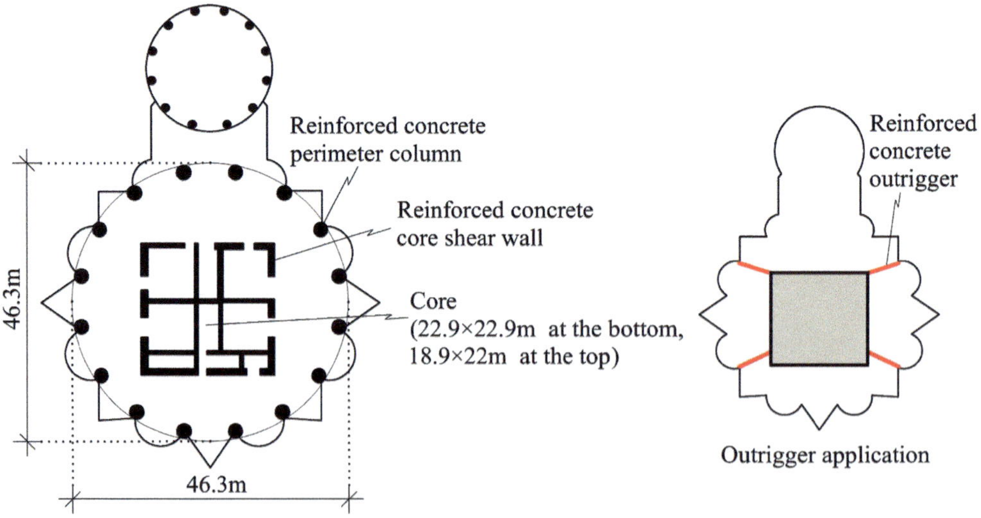

FIGURE 3.31 The Petronas Twin Towers, Kuala Lumpur, Malaysia, 1998

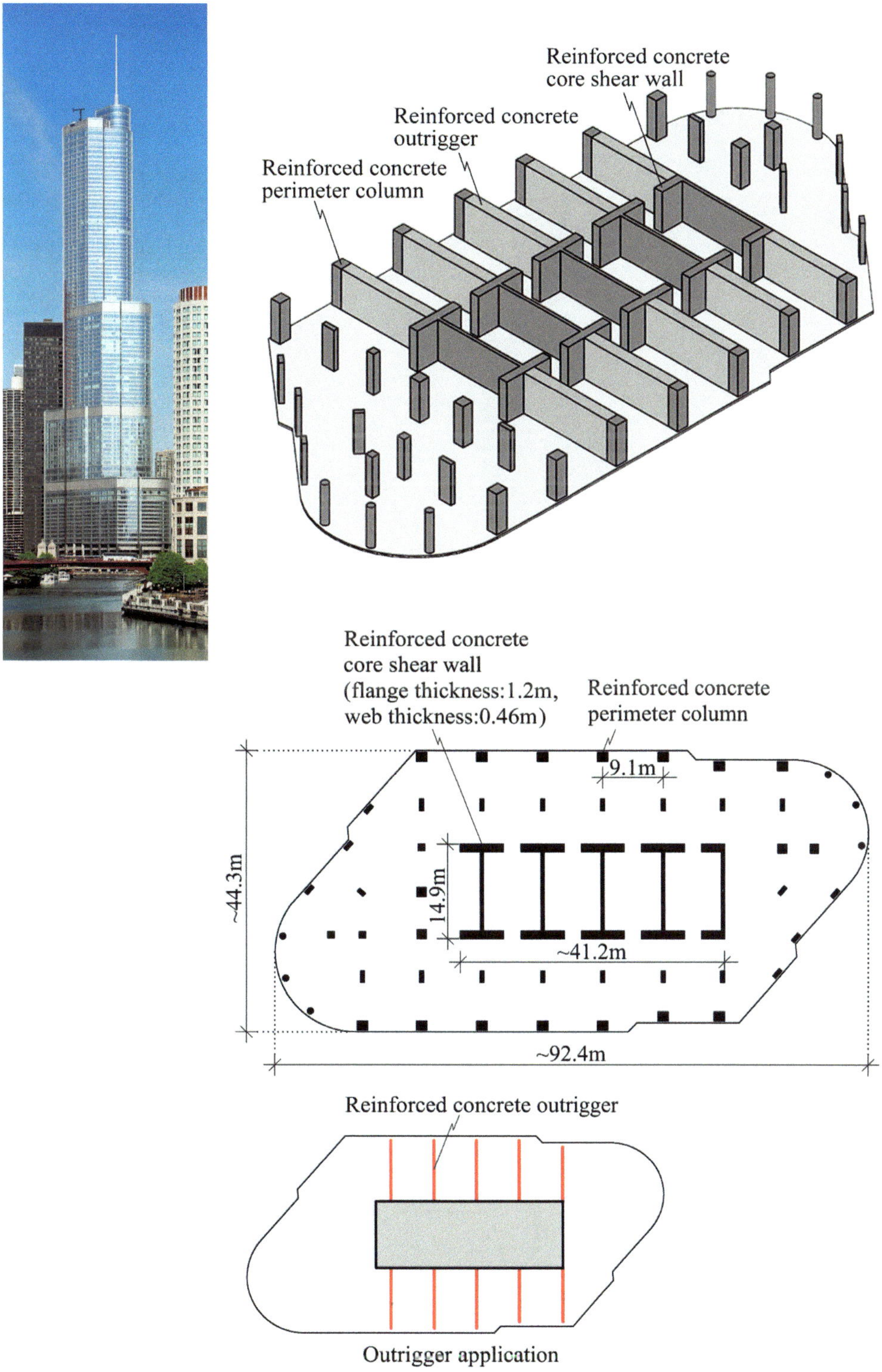

FIGURE 3.32 Trump International Hotel & Tower, Chicago, USA, 2009

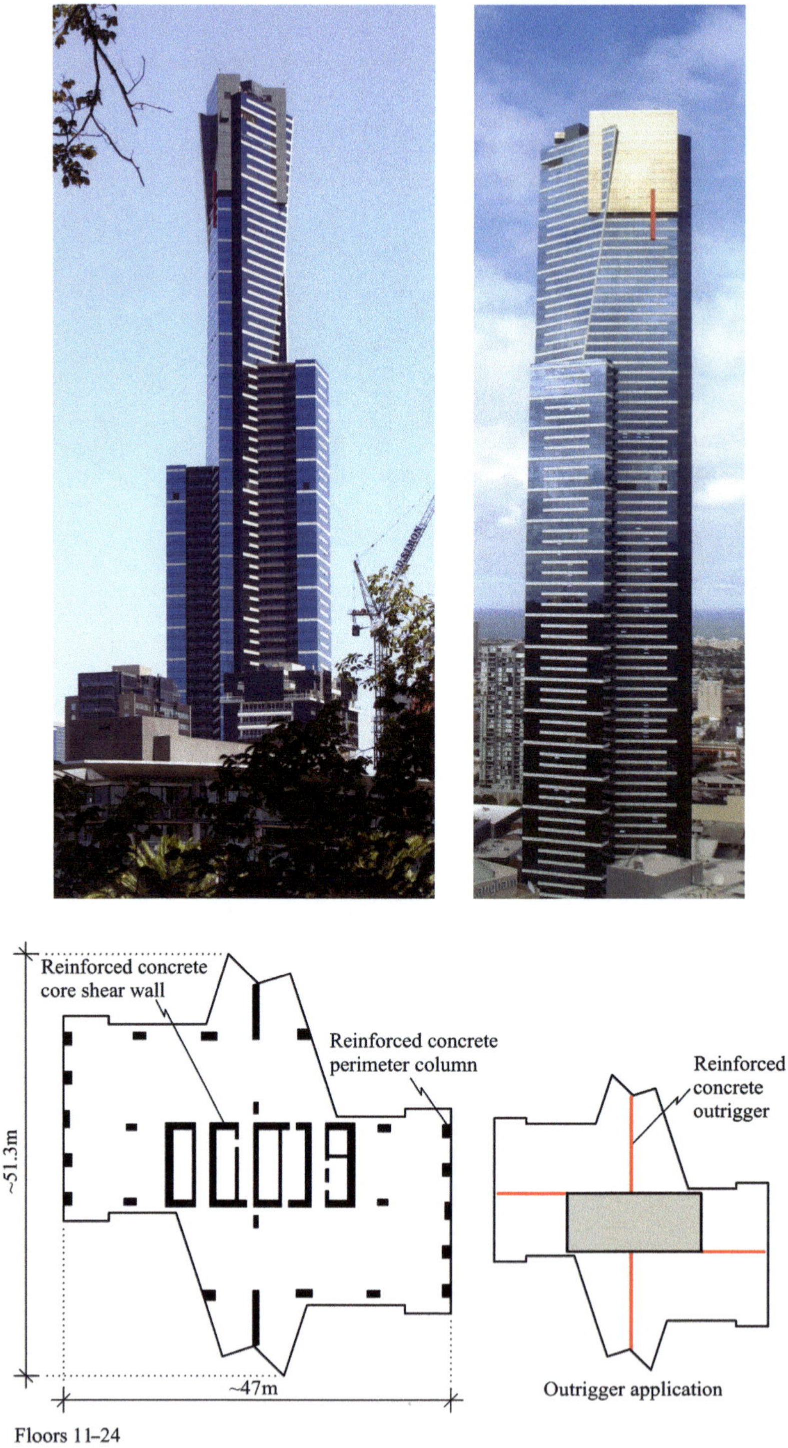

FIGURE 3.33 Eureka Tower, Melbourne, Australia, 2006
(photos courtesy of David Randerson)

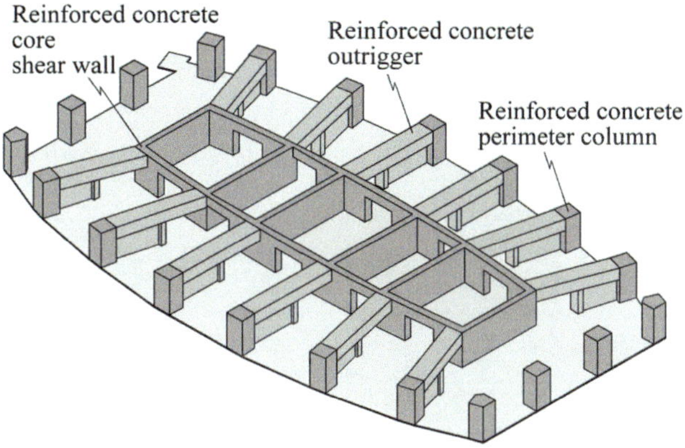

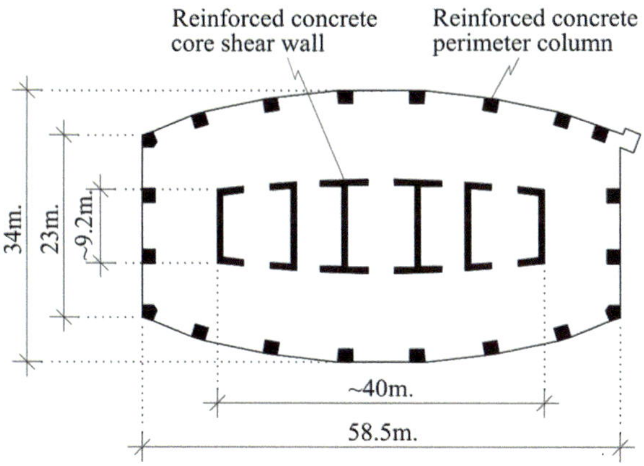

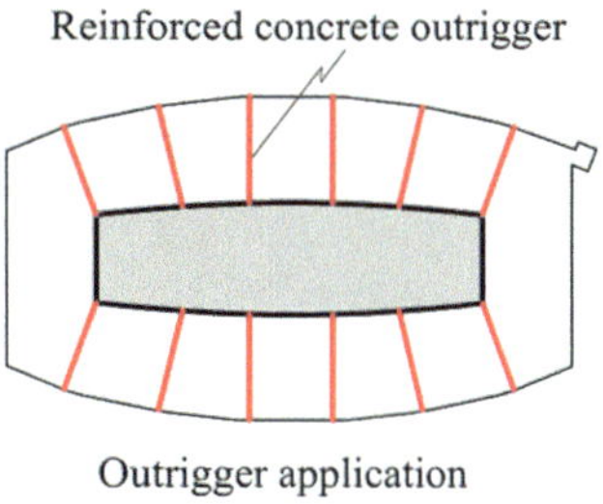

FIGURE 3.34 Plaza 66, Shanghai, China, 2001

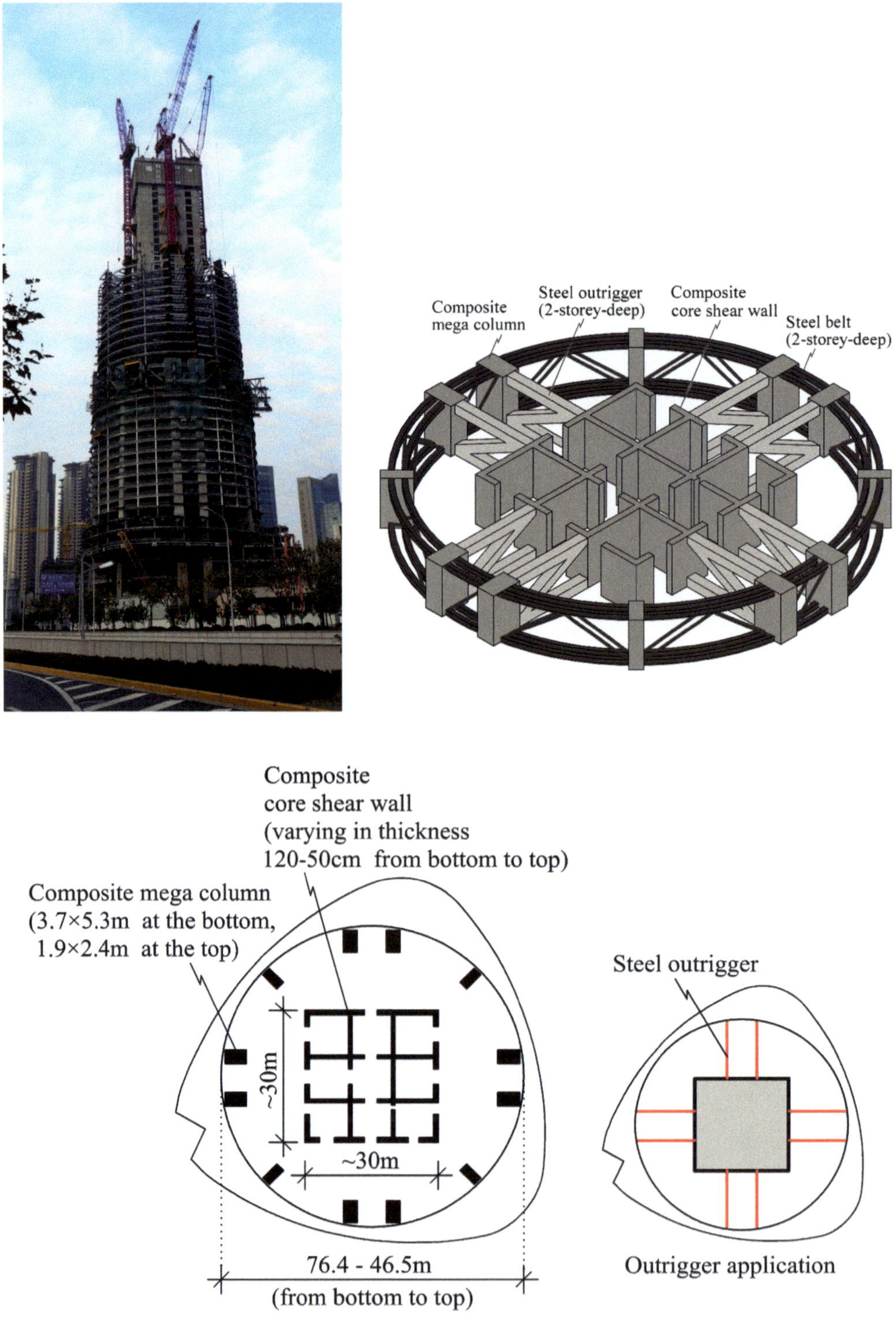

FIGURE 3.35 Shanghai Tower, Shanghai, China, under construction

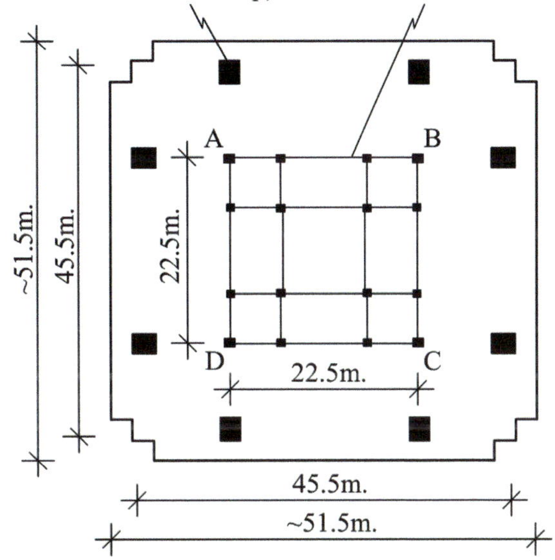

Floor 42

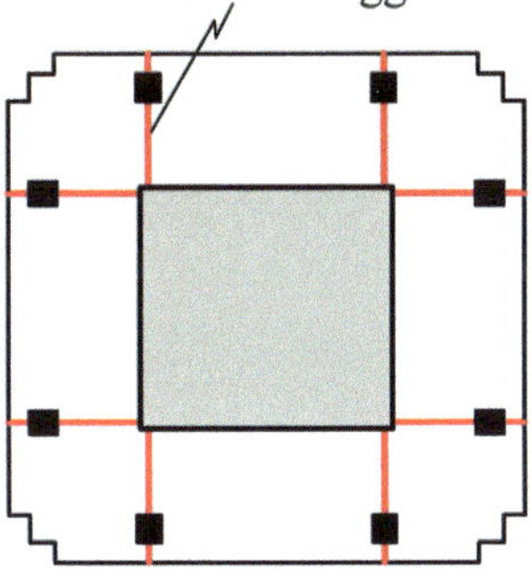

Outrigger application

FIGURE 3.36 Taipei 101, Taipei, Taiwan, 2004

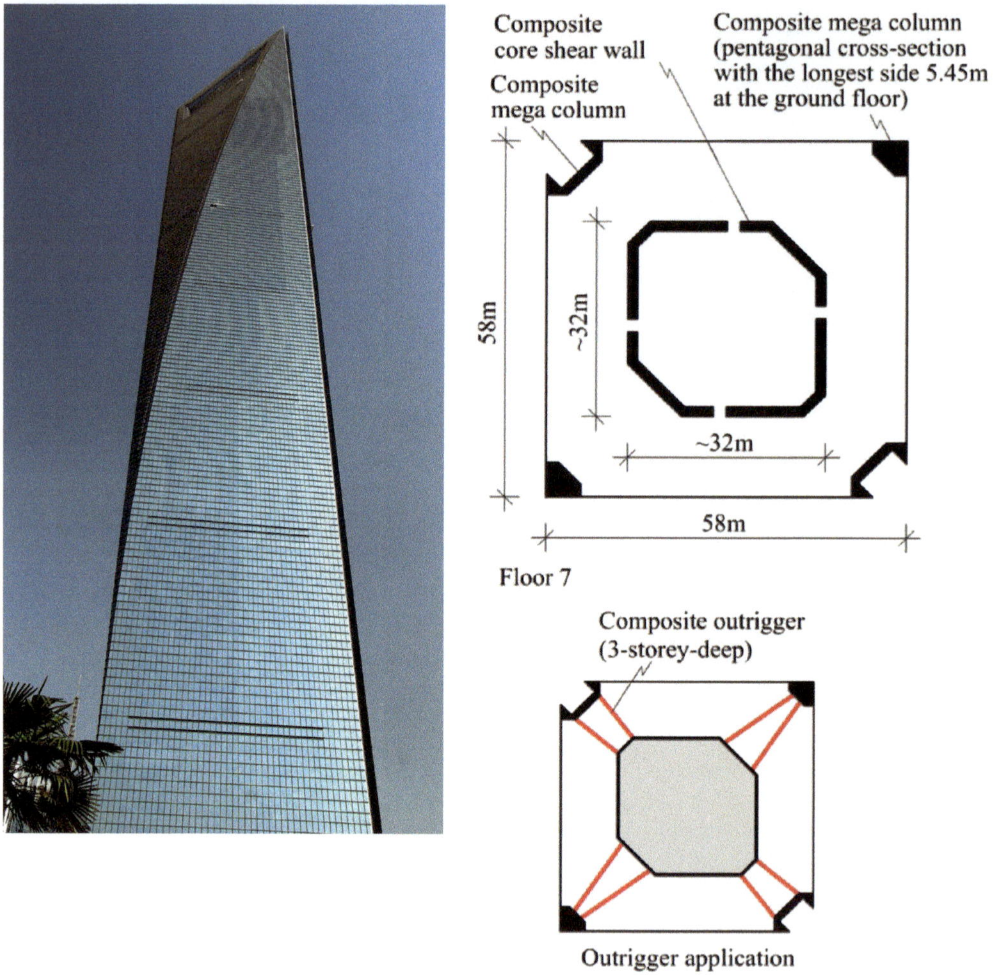

FIGURE 3.37 Shanghai World Financial Center, Shanghai, China, 2008
(photo courtesy of Niels Jakob Darger)

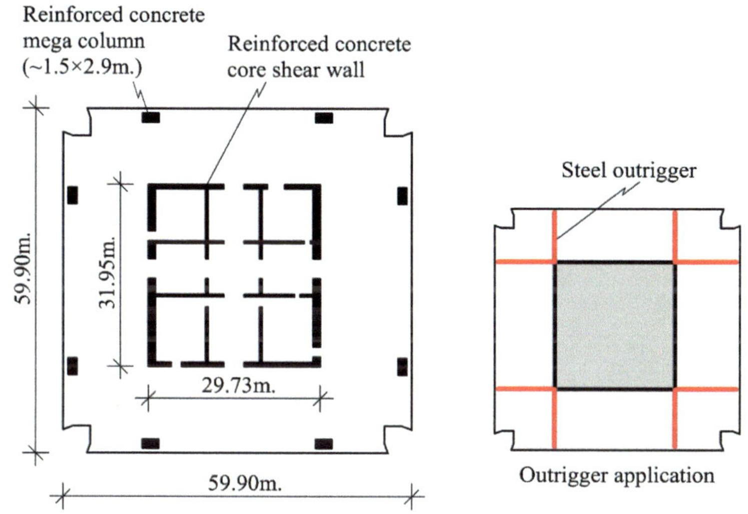

Floor 75

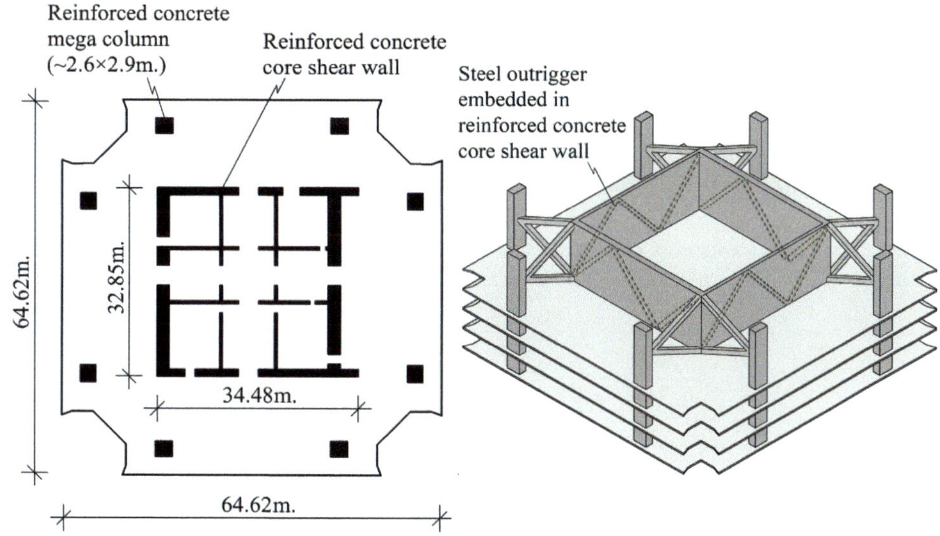

Floor 15

FIGURE 3.38 International Commerce Centre (ICC), Hong Kong, China, 2010 *(photo on left and drawing on right courtesy of Sun Hung Kai Properties)*

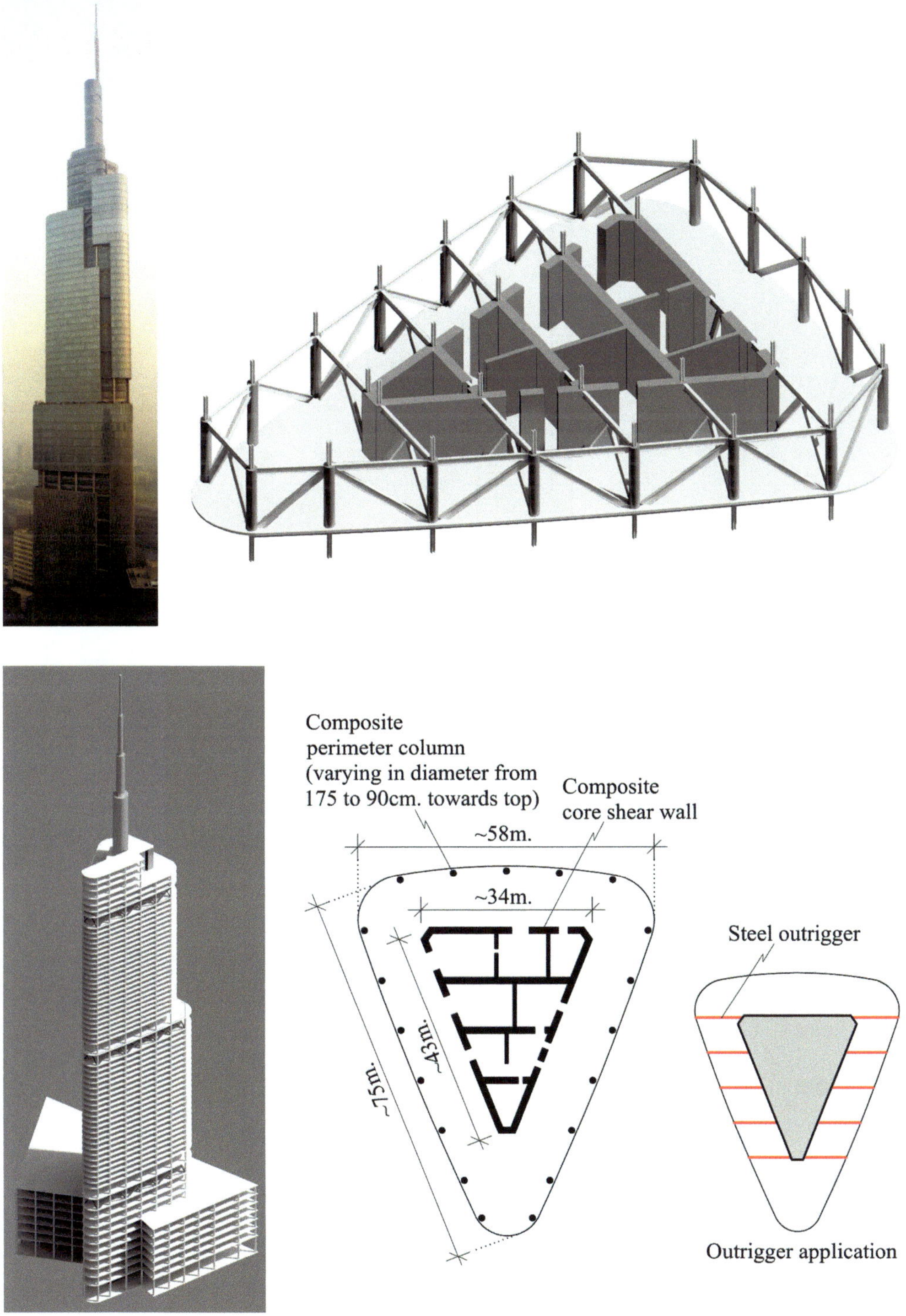

FIGURE 3.39 Zifeng Tower, Nanjing, China, 2010
(top right and bottom left drawings courtesy of Ramazan Sari)

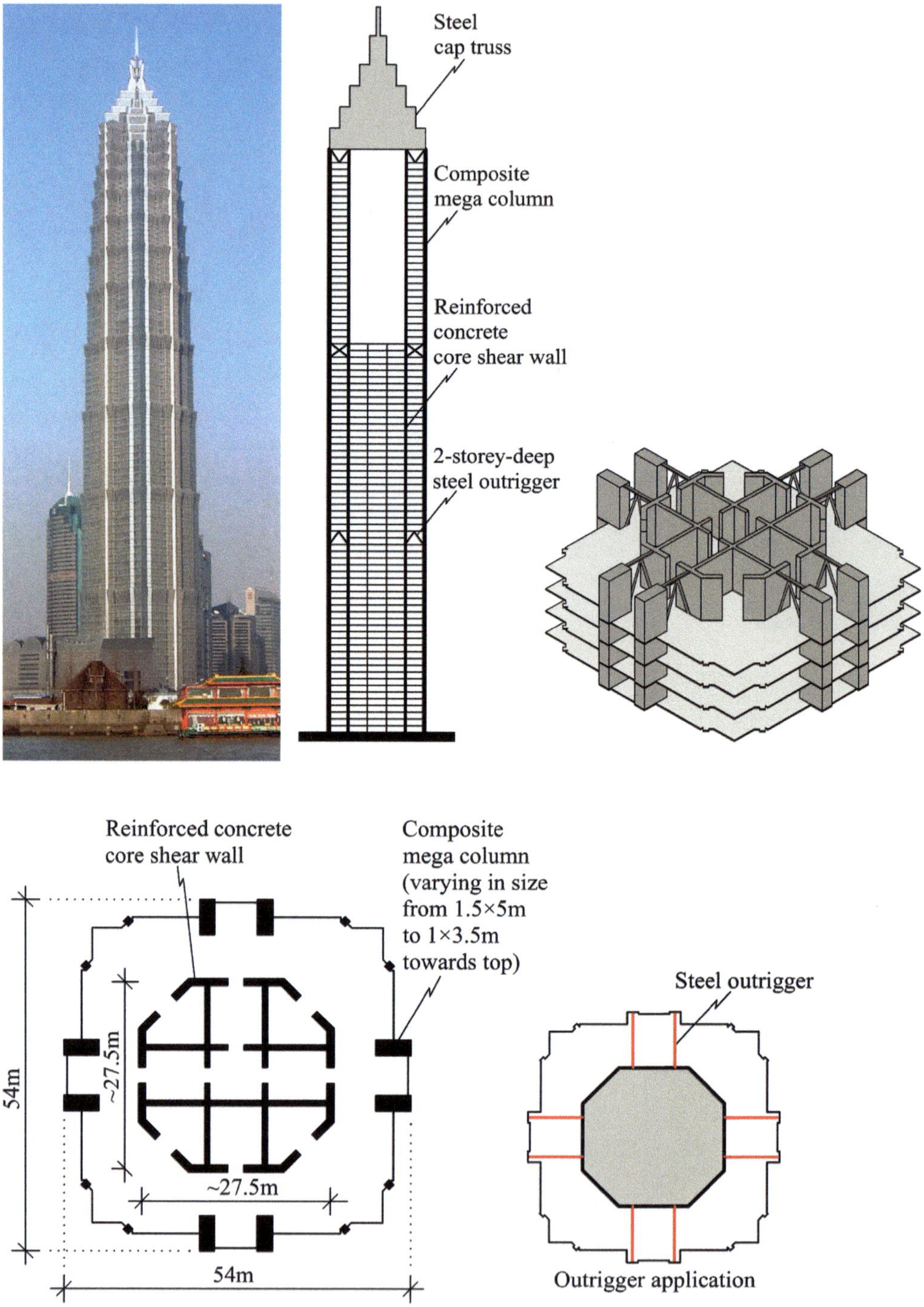

FIGURE 3.40 Jin Mao Building, Shanghai, China, 1999
(photo courtesy of Wilfried Blümler)

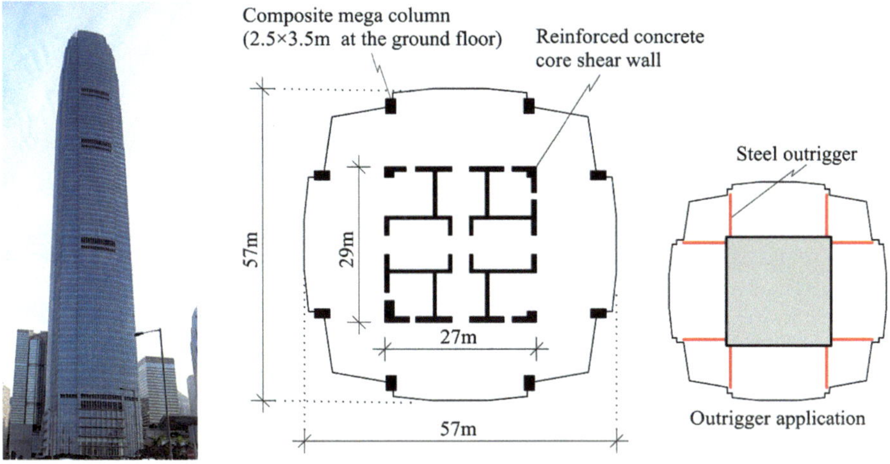

FIGURE 3.41 Two International Finance Centre, Hong Kong, China, 2003 *(photo courtesy of Niels Jakob Darger)*

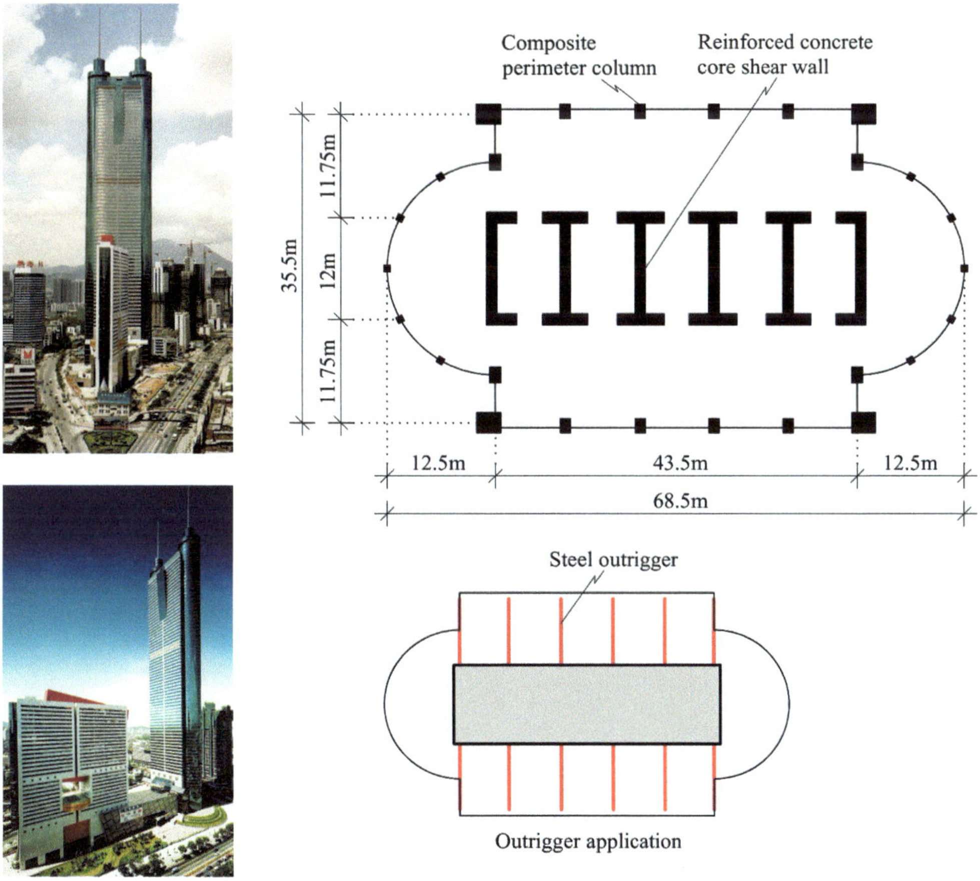

FIGURE 3.42 Shun Hing Square, Shenzhen, China, 1996 *(photos courtesy of Derek Forbes)*

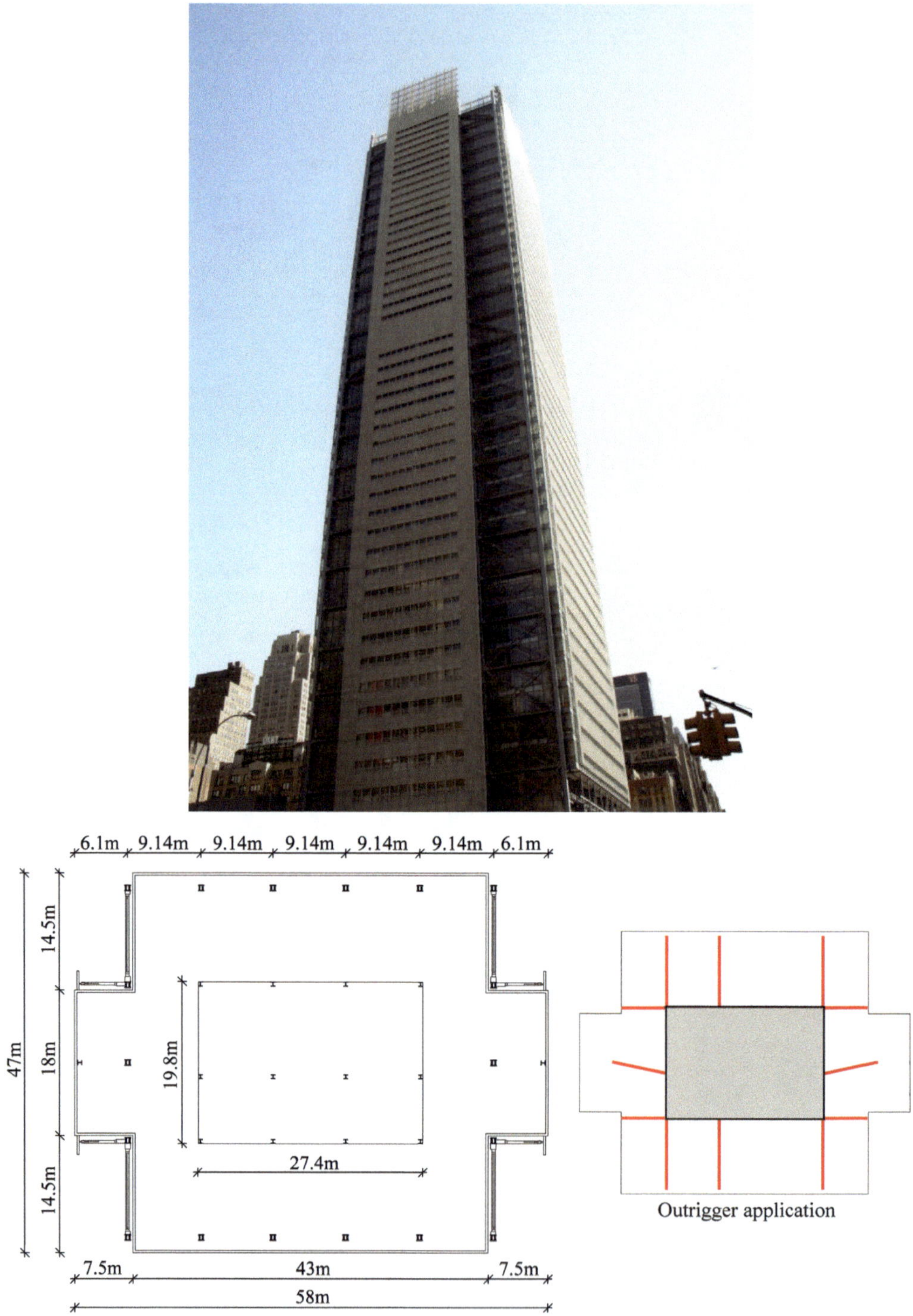

FIGURE 3.43 New York Times Tower, New York, USA, 2007
(photo courtesy of Antony Wood/CTBUH and plan courtesy of Ilkay Guryay)

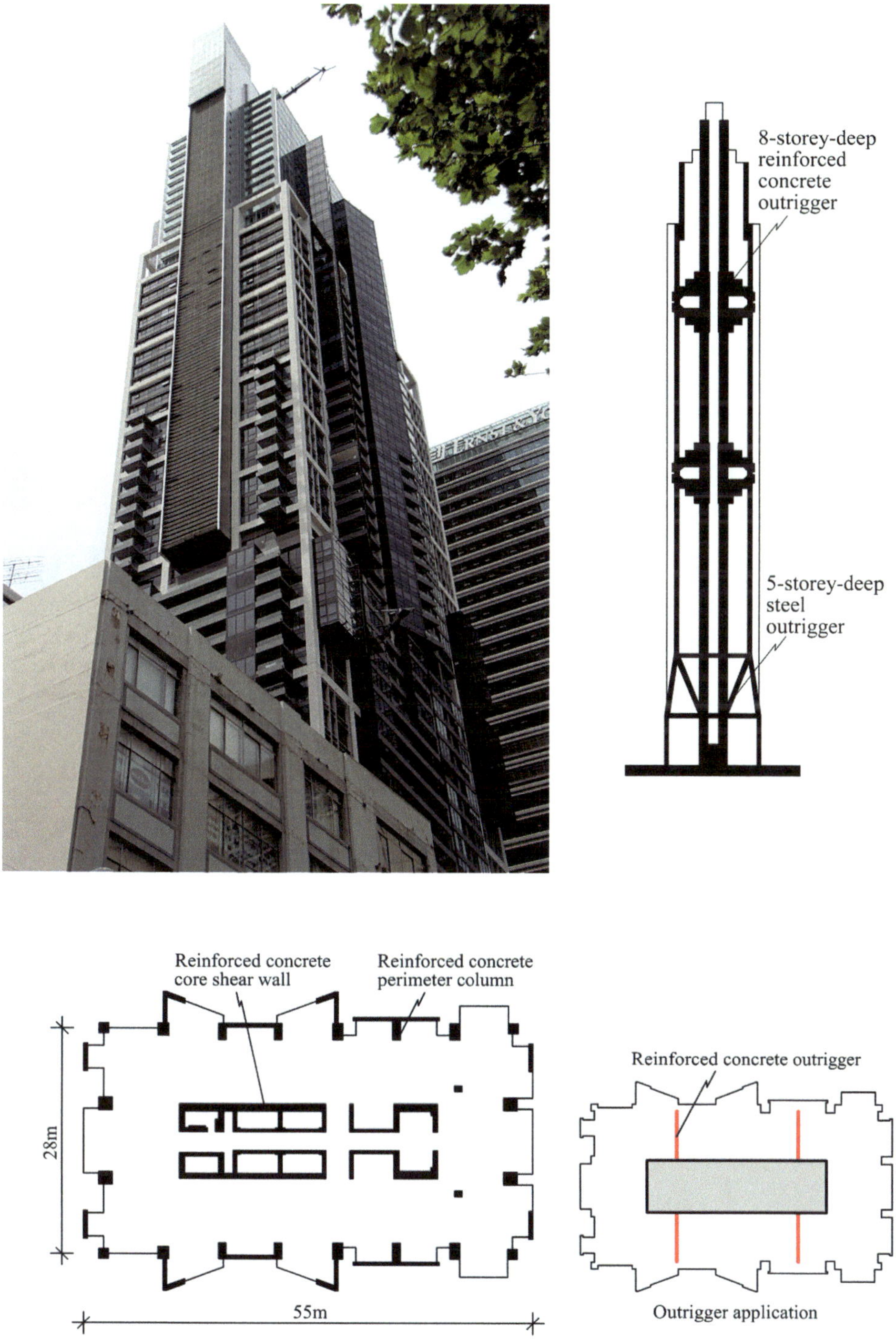

FIGURE 3.44 World Tower, Sydney, Australia, 2004
(photo courtesy of Niels Jakob Darger)

3.9.1 The behaviour of outriggered frame systems

When outriggers cantilevered from the core are connected rigidly to the perimeter (exterior) columns (Figure 3.45a), these columns are subject to additional bending moments and axial forces transferred from the outriggers, and the system cannot completely benefit from the moment carrying capacity of the shear core. On the other hand, when outriggers are connected by hinges to the perimeter columns (Figure 3.45b), by blocking the transfer of the bending moment from the outriggers to the columns, the column axial load capacity is increased and the system completely benefits from the moment carrying capacity of the shear core. For this reason, hinged connections between outriggers and perimeter columns increase the efficiency of the system by maximising the utilisation of not only the moment resisting capacity of the shear core but also the axial capacity of the columns.

An analysis is given below within a framework of various assumptions for outriggered frame systems under uniformly distributed lateral loads, where outriggers are located at one or two levels throughout the height of the building.

In a simplified analytical model, the behaviour of an outriggered frame system under lateral loads (Figure 3.46) can be separated into two as a vertical cantilever core under lateral loads and as the same core with restoring moment created by the outrigger's levering effect (Figure 3.47).

The outrigger transfers the restoring moment to the core, acting as a fulcrum, with the levering effect restrained by the perimeter columns and resists the rotation of the

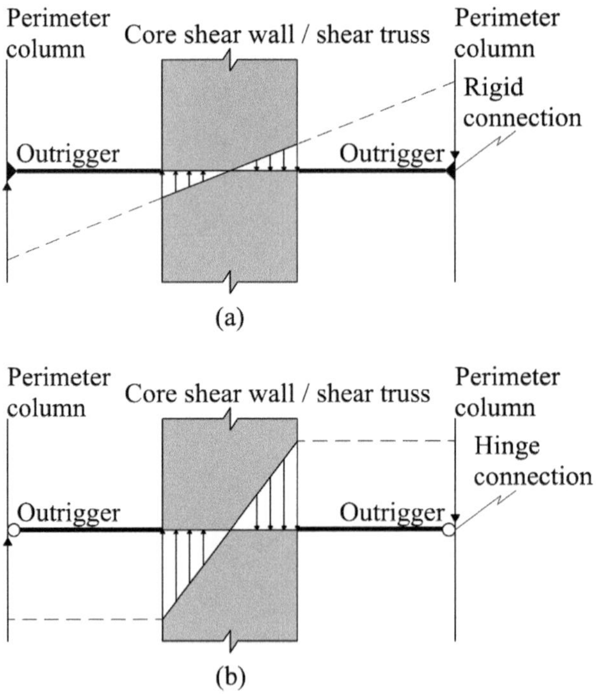

FIGURE 3.45 Outrigger to perimeter column connections: (a) rigid connection, (b) hinged connection

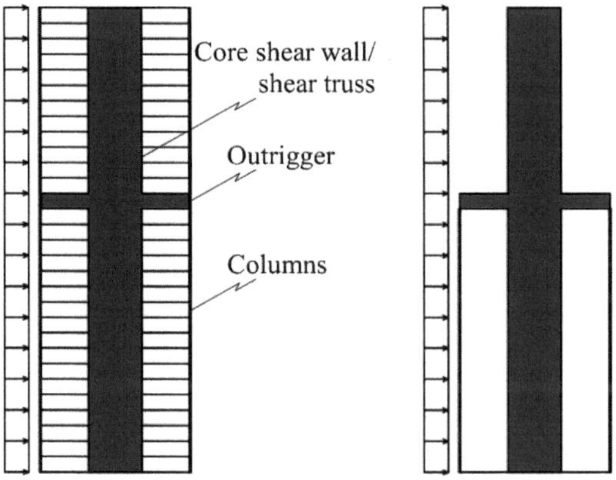

FIGURE 3.46 Outriggered frame system under lateral loads and analytical model

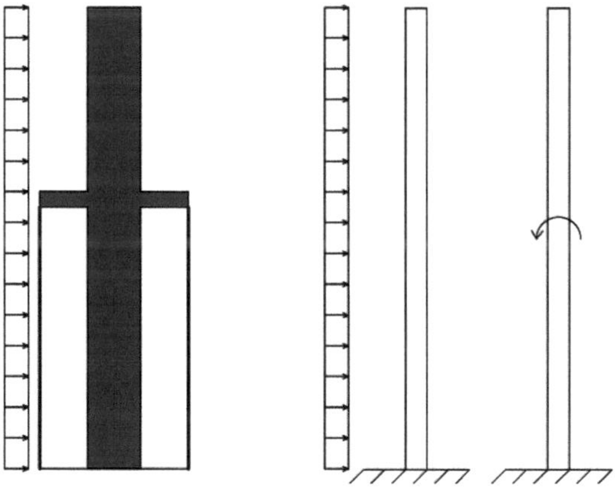

FIGURE 3.47 Superposition of analytical model

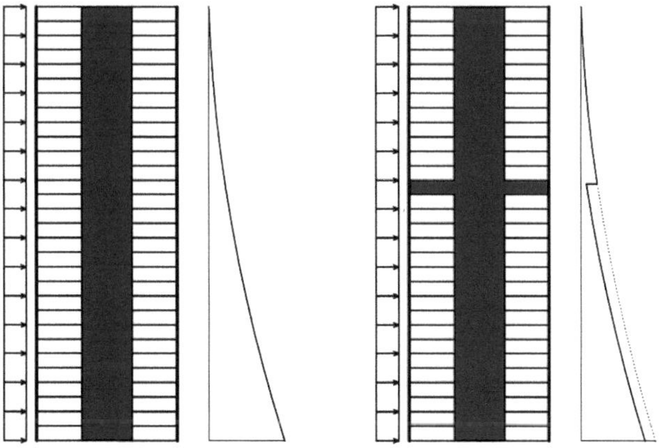

FIGURE 3.48 Diagram of the effect of the outrigger on the moment

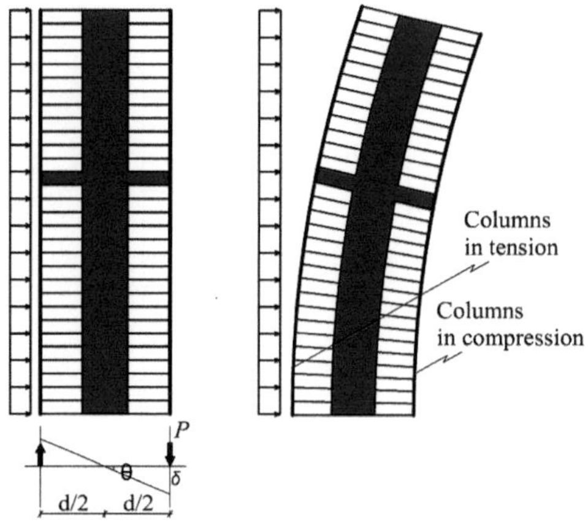

FIGURE 3.49 Axial deformation of the perimeter columns on two facades perpendicular to the bending direction

core under lateral loads. Thus, by reducing the rotation of the core, both the bending moment of the core (Figure 3.48) and the lateral drift at the top of the building are reduced. Columns on two facades perpendicular to the bending direction are subjected to axial tension or compression, and so are elongated on one side and contracted on the other (Figure 3.49).

The behaviour of outriggered frame systems is analysed below, taking the lateral drift at the building top as a basis, in cases where the outriggers are located throughout the height of the building at one or two levels. Assuming that the outriggers are so stiff that rotation in the outriggers due to axial deformation of the columns and the rotation of the core at the same level under lateral loads are equal, the restoring moment created by the outrigger and the lateral drift of the outrigger are obtained with the help of compatibility equations. Below, an approximate analysis is presented, assuming uniform columns, uniform core and uniform outriggers, hoping to be helpful to create a rough estimation in preliminary design stage.

The analysis has been made with the following assumptions:

- The core is a vertical cantilever rigidly fixed at the base and rigid against shear.
- Outriggers are rigidly fixed to the core, have hinged connections to the perimeter columns to induce axial forces only, and are rigid against shear and flexure.
- The cross-sectional areas of the columns are constant from the top outrigger down to foundation and the moment of inertia of the core is constant throughout the building height.
- The lateral load on the building is constant throughout the building height.
- The structure is linearly elastic.

Outriggers can be represented by an equivalent spring of rotational stiffness K at the core. The rotational stiffness of this spring (the moment per unit of rotation) for a couple of columns (working on opposite sides perpendicular to the bending direction) under an axial load p_i:

$$K_i = p_i d = \frac{A_i E d^2}{2L}$$

where E is the modulus of elasticity.

Considering all the couple of columns (in elongation and contraction) on the two facades perpendicular to the bending direction:

$$K = \sum_{i=1}^{n} K_i = \frac{Ed^2}{2L} \sum_{i=1}^{n} A_i = \frac{AEd^2}{2L}$$

$$K = \frac{AEd^2}{2L}$$

The rotational stiffness of outriggers at distance x from the top of the structure is:

$$K = \frac{AEd^2}{2(L - x)}$$

The function of the rotational stiffness of the spring, representing the effect of the outrigger on the core, shows that the rotational stiffness of the outrigger is directly proportional to the distance of its location from the top of the structure.

3.9.1.1 The optimum location of a single outrigger level

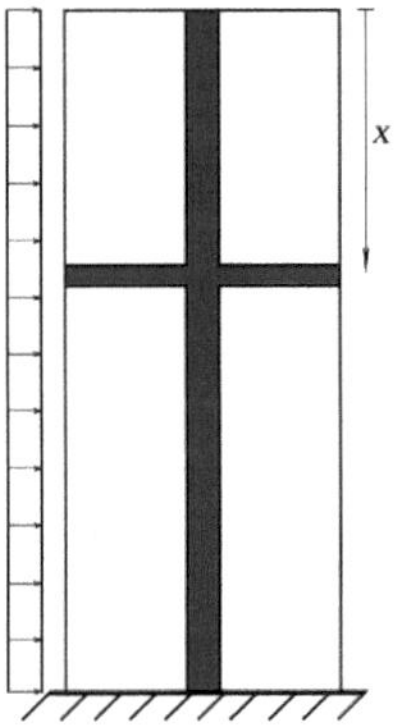

FIGURE 3.50 Location of a single outrigger level at distance x from the top

For the location of a single outrigger level at distance x from the top of the structure (Figure 3.50), the lateral drift at distance x from the top of the structure is:

$$y_{x=0} = \frac{wL^4}{8EI} - \frac{M_x}{2EI}(L^2 - x^2)$$

and the restoring effect of the outrigger on the lateral drift at the top of the structure is:

$$y_{rx=0} = \frac{M_x}{2EI}(L^2 - x^2)$$

where EI is the flexural rigidity.

M_x must be found in the above equations. This is done using the rotation equations. The rotation of the core at the level where the outrigger is located is:

$$\theta_{cx} - \theta_{rx} = M_x / K_x$$

$$\frac{w}{6EI}(L^3 - x^3) - \frac{M_x}{EI}(L - x) = \frac{M_x}{K_x}$$

From the above equation, M_x is found.

$$M_x = \frac{w/(6EI)}{\frac{L}{EI} - \frac{x}{EI} + \frac{1}{K_x}}(L^3 - x^3) = \frac{wL^2}{6EIC}(x^2 + x + 1)$$

In the equation for the outrigger effect on the lateral drift at the top of the structure, assuming that

$$C = \frac{1}{EI} + \frac{2}{AEd^2}$$

then,

$$y_{rx=0} = \frac{w}{12(EI)^2\,C}(L^3 - x^3)(L + x)$$

The lateral drift at the top of the structure is:

$$y_{x=0} = \frac{wL^4}{8EI} - \frac{w}{12(EI)^2\,C}(L^3 - x^3)(L + x)$$

The location of the outrigger level that has the greatest effect on the lateral drift at the top of the structure is the location where the $y_{x=0}$ function has the minimum, or $y_{rx=0}$ function has the maximum value. Therefore, the optimum location of a single outrigger level is obtained by differentiating the $y_{rx=0}$ function with respect to x, and equating to zero.

$$x = 0.455L$$

In the equation for the lateral drift at the top of the structure, for x=0.455L, the lateral drift at the top of the structure is:

$$y = \frac{wL^4}{8EI} - \frac{wL^4}{12(EI)^2\,C}1.32$$

and the restoring moment of the outrigger is:

$$M = \frac{wL^2}{6EIC}1.66$$

3.9.1.2 The optimum location of two outrigger levels

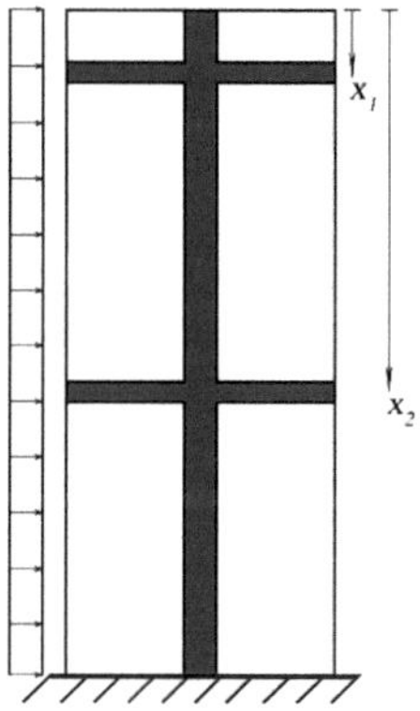

FIGURE 3.51 Location of two outrigger levels at distances x_1 and x_2 from the top

For the location of two outrigger levels at distances x_1 and x_2 from the top of the structure, the lateral drift at the top of the structure is (Figure 3.51):

$$y_{x=0}\ \frac{wL^4}{8EI} - \frac{1}{2EI}\left[M_1\,(L^2 - x_1^2) + M_2\,(L^2 - x_2^2)\right]$$

and the restoring effect of the outriggers on the lateral drift at the top of the structure is:

$$y_{rx=0} = \frac{1}{2EI}\left[M_1\,(L^2 - x_1^2) + M_2\,(L^2 - x_2^2)\right]$$

M_1 and M_2 must be found in the above equations. This is done using the rotation equations.

The rotations of the core at the levels where the outriggers are located are:

$$\theta_{x1} = \frac{w}{6EI}(L^3 - x_1^3) - \frac{M_1}{EI}(L - x_1) - \frac{M_2}{EI}(L - x_2)$$

$$\theta_{x2} = \frac{w}{6EI}(L^3 - x_2^3) - \frac{M_1}{EI}(L - x_2) - \frac{M_2}{EI}(L - x_2)$$

On the other hand, since

$$\theta = \frac{M}{K}$$

$$\theta_{x1} = \frac{M_1}{\dfrac{AEd^2}{2(L - x_1)}} + \frac{M_2}{\dfrac{AEd^2}{2(L - x_2)}} = \frac{2M_1(L - x_1)}{AEd^2} + \frac{2M_2(L - x_2)}{AEd^2}$$

$$\theta_{x2} = \frac{M_1}{\dfrac{AEd^2}{2(L - x_2)}} + \frac{M_2}{\dfrac{AEd^2}{2(L - x_2)}} = \frac{2M_1(L - x_2)}{AEd^2} + \frac{2M_2(L - x_2)}{AEd^2}$$

Assuming that:

$$C = \frac{1}{EI} + \frac{2}{AEd^2}$$

From the above equations, M_1 and M_2 can be found.

$$M_1 = \frac{w}{6EIC}\frac{(x_2^3 - x_1^3)}{(x_2 - x_1)} = \frac{w}{6EIC}(x_1^2 + x_1 x_2 + x_2^2)$$

$$M_2 = \frac{w}{6EIC}(L - x_1)(L + x_1 + x_2) = \frac{w}{6EIC}(L^2 + Lx_2 - x_1 x_2 - x_1^2)$$

The lateral drift at the top of the structure is:

$$y_{x=0} = \frac{wL^4}{8EI} - \frac{w}{12(EI)^2 C}\left[(x_1^2 + x_1 x_2 + x_2^2)(L^2 - x_1^2) + (L^2 + Lx_2 - x_1 x_2 - x_1^2)(L^2 - x_2^2)\right]$$

The location of the outrigger levels that have the greatest effect on the lateral drift at the top of the structure are the locations where the $y_{rx=0}$ function has the maximum values. Therefore, the optimum locations for two outrigger levels are obtained by differentiating the $y_{rx=0}$ function with respect to x_1 and x_2 and equating to zero.

$$x_1 = 0.31L$$

$$x_2 = 0.69L$$

The lateral drift at the top of the structure is:

$$y = \frac{wL^4}{8EI} - \frac{wL^4}{12(EI)^2 C}1.44$$

and the restoring moments of the outriggers are:

$$M_1 = \frac{wL^2}{6EIC}0.79 \quad M_2 = \frac{wL^2}{6EIC}1.38$$

3.9.1.3 Two outrigger levels, one at the top of the structure and the other at the optimum location

In the equation of the restoring effect of two outrigger levels on the lateral drift at the top of the structure, when $x_1=0$, the optimum location x_2 is obtained by differentiating the $y_{rx=0}$ function with respect to x_2 and equating to zero.

$$x_2 = 0.5774L$$

In the equation for the lateral drift at the top of the structure for two outrigger levels, when $x_1=0$, and $x_2=0.5774L$, the lateral drift at the top of the structure is:

$$y = \frac{wL^4}{8EI} - \frac{wL^4}{12(EI)^2 C} 1.38$$

and the restoring moments of the outriggers are:

$$M_1 = \frac{wL^2}{6EIC} 0.33 \quad M_2 = \frac{wL^2}{6EIC} 1.58$$

3.9.1.4 The lateral drift at the top of the structure when a single outrigger level is located at the top of the structure

The equation for the lateral drift at the top of the structure for a single outrigger level is used, and x is taken as zero (x=0) since the outrigger level is located at the top of the structure.

The lateral drift at the top of the structure is:

$$y_{y=0} = \frac{wL^4}{8EI} - \frac{wL^4}{12(EI)^2 C}$$

and the restoring moment of the outrigger is:

$$M = \frac{wL^2}{6EIC}$$

3.9.2 Evaluation of outriggered frame systems

The optimum locations, lateral drift at the building top and restoring moments of outriggered frame systems are summarised in Table 3.2.

According to Table 3.2:

1. Assessing the reduction of lateral drift at the top of the structure due to the addition of outriggers to a shear-frame (shear trussed/braced or shear walled frame) system,
   a. for a single outrigger level at the top of the structure, the lateral drift at the top of the structure is reduced by 67 per cent/(EIC)
   b. for a single outrigger level at the optimum location, the lateral drift at the top of the structure is reduced by 88 per cent/(EIC),
   c. for two outrigger levels, one at the top of the structure and the other at the optimum location, the lateral drift at the top of the structure is reduced by 92 per cent/(EIC),
   d. for two outrigger levels at the optimum locations, the lateral drift at the top of the structure is reduced by 96 per cent/(EIC),

TABLE 3.2 Evaluation of outriggered frame systems

			Restoring moment
Single outrigger level at the top of the structure	x=0	$y_{x=0} = \frac{wL^4}{8EI} - \frac{wL^4}{12(EI)^2 C}$	$M = \frac{wL^2}{6EIC}$
Single outrigger level at the optimum location	x=0.455 L	$y = \frac{wL^4}{8EI} - \frac{wL^4}{12(EI)^2 C} 1.32$	$M = \frac{wL^2}{6EIC} 1.66$
Two outrigger levels, one at the top of the structure	x_1=0 x_2=0.5774 L	$y = \frac{wL^4}{8EI} - \frac{wL^4}{12(EI)^2 C} 1.38$	$M_1 = \frac{wL^2}{6EIC} 0.33$ $M_2 = \frac{wL^2}{6EIC} 1.58$
Two outrigger levels at the optimum locations	x_1=0.31 L x_2=0.69 L	$y = \frac{wL^4}{8EI} - \frac{wL^4}{12(EI)^2 C} 1.44$	$M_1 = \frac{wL^2}{6EIC} 0.79$ $M_2 = \frac{wL^2}{6EIC} 1.38$

where EI is the flexural rigidity, and C is the constant defined by the following equation:

$$C = \frac{1}{EI} + \frac{2}{AEd^2}$$

2. Assessing the contribution of outriggers to the reduction in lateral drift at the top of the structure,
   a. the contribution of a single outrigger level at its optimum location is 32 per cent higher than the contribution from a single outrigger level located at the top of the structure.
   b. the contribution of two outrigger levels at their optimum locations is 12 per cent higher than the contribution of a single outrigger level at its optimum location.
   c. two outrigger levels, one at the top of the building structure and the other at the optimum location, contribute 6 per cent more than that of a single outrigger level at its optimum location.

In this case, it is debatable whether the reduction of 6 per cent of the lateral drift justifies the economic cost of adding a second outrigger level at the top of the structure.

3. 64 per cent of the total restoring moment of two outrigger levels at the optimum locations comes from the lower outrigger; 83 per cent of the total restoring moment of two outrigger levels, where one is at the top of the structure and the

other is at the optimum location, comes from the lower outrigger. The outrigger that is closest to the base of the structure provides the most restoring moment, and its share in the total restoring moment is reduced when the number of outriggers is increased and it is placed in the optimum location.

3.10 Tube systems

The tube system was innovated in the early 1960s by the famous structural engineer Fazlur Rahman Khan who is considered the "father of tubular design" (Weingardt, 2011). The tube system can be likened to a system in which a hollow box column is cantilevering from the ground, and so the building exterior exhibits a tubular behaviour against lateral loads. This system is evolved from the rigid frame system and can be defined as a three-dimensional rigid frame having the capability of resisting all lateral loads with the facade structure. The tube system was used for the first time as the framed-tube system in the 43-storey, 120 m high The Plaza on Dewitt (formerly Dewitt-Chestnut Apartments) (Chicago, 1966) (Figure 3.52) by Fazlur Rahman Khan.

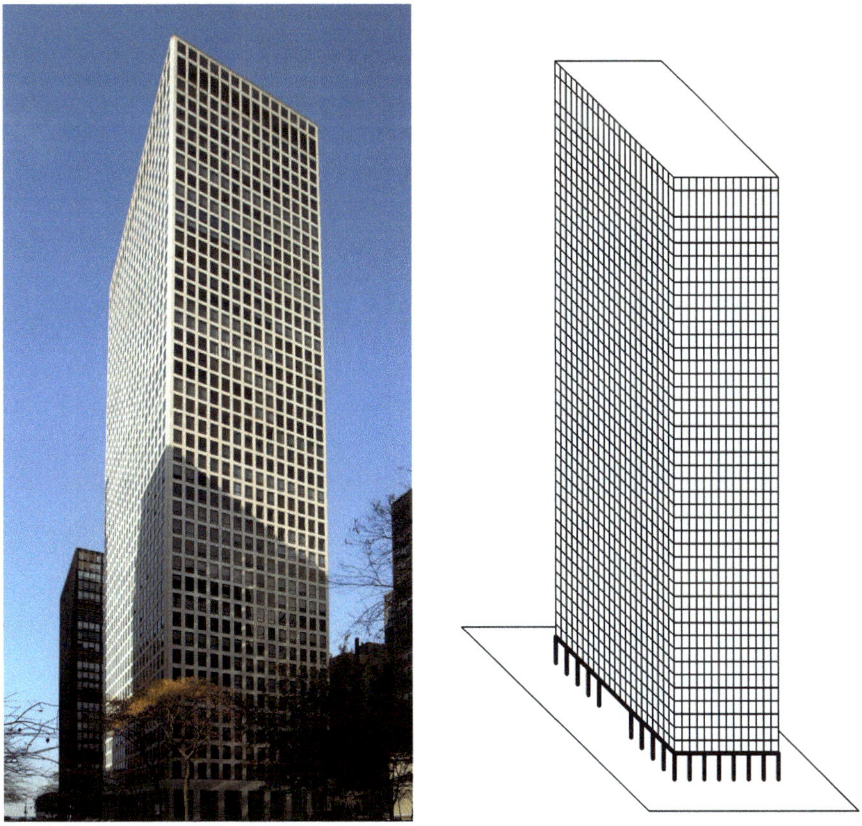

FIGURE 3.52 The Plaza on Dewitt, Chicago, USA, 1966
(photo courtesy of Marshall Gerometta/CTBUH)

In tubular design, the rigidity of the structural system against lateral loads can be increased with solutions such as:

- closer spacing of the perimeter columns
- increasing the depth of the spandrel beams connected to the perimeter columns
- adding shear trusses/braces or shear walls to the core
- adding an inner tube in place of the core (tube-in-tube)
- adding a truss (multi-storey braces) to the building exterior (trussed-tube)
- combining more than one tube (bundled-tube).

In tube systems, the tube formed around the building exterior is designed to resist all lateral and vertical loads. If there is a structural core in the interior of the building, it is assumed to support some part of the vertical loads. Adding a second tube instead of a core can increase the stiffness of the structural system to support some part of the vertical and lateral loads.

As well as its structural efficiency, in a tube system it increases the net usable area of the building while reducing the dimensions of the structural elements in the core, thanks to the tubular exterior frame supporting the entire lateral load. Tube systems can be used in several geometrical forms like rectangular, square, triangular, circular and even free-forms in the plan (Figure 3.53).

Tube systems efficiently and economically provide sufficient stiffness to resist wind and earthquake induced lateral loads in buildings of more than 40 storeys.
Tube systems can be divided into three types:

- framed-tube systems
- trussed-tube systems
- bundled-tube systems.

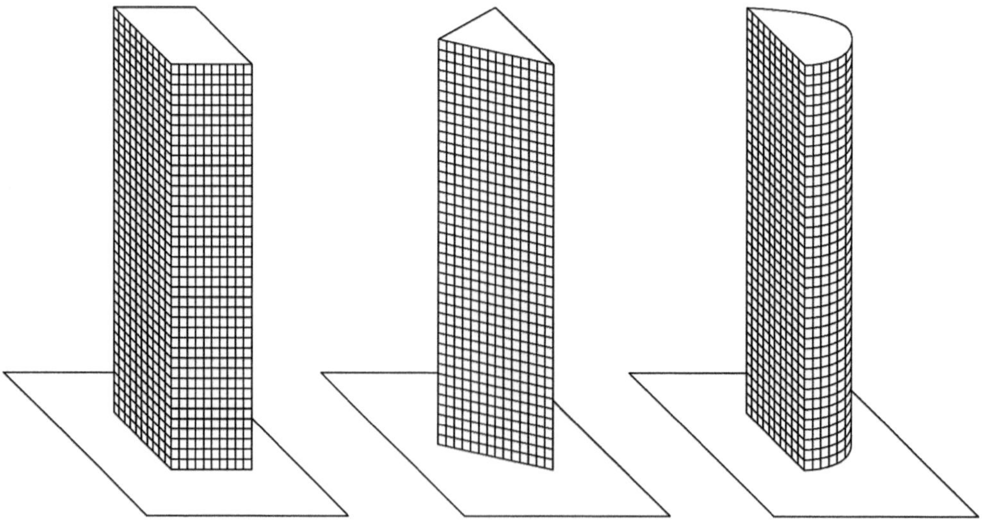

FIGURE 3.53 Some forms of tube systems

3.10.1 Framed-tube systems

The framed-tube systems, which constitute the basis of tube systems, can be described as having evolved from rigid frame systems and are alternative to shear frame systems. The outstanding structural engineer Fazlur Rahman Khan innovated the framed-tube system.

The most significant feature of the system, also known as the "vierendeel tube system" or "perforated tube system", is the closely spaced perimeter/exterior columns, which are usually spaced at 1.5 to 4.5 m centres, connected by deep spandrel beams at floor levels. If there is a need to increase the column spacing, in order to secure the behaviour of the framed-tube system, it is necessary to increase the dimensions of the perimeter columns and spandrel beams.

The dimensions and spacing of the columns and the flexural rigidity of the spandrel beams directly affect the tubular behaviour of the framed-tube system. In the framed-tube system, pure tubular cantilever behaviour cannot be fully achieved because of the flexibility of the spandrel beams so that there can be slight bending deformation while transferring the shear forces to the columns. The real behaviour of the system is between the behaviour of a vertical cantilever and that of a frame. Limited flexural and shear rigidity (flexibility) of the spandrel beams results in bending deformation, so the axial stresses in the corner perimeter columns increase while they decrease in the inner perimeter columns. In this way, the distribution of axial compressive and tensile stresses formed in the perimeter columns in response to the lateral loads cannot be linear (Figure 3.54). This phenomenon is known as "shear lag", which depends

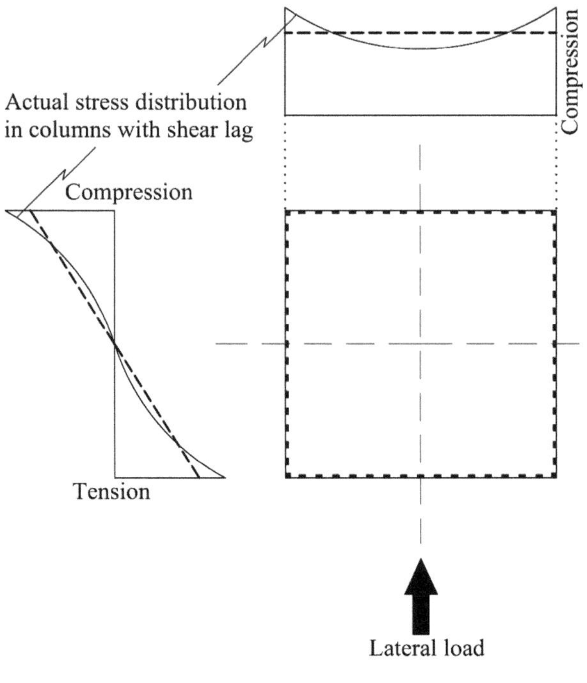

FIGURE 3.54 The distribution of tension and shear lag in perimeter columns in a framed-tube system

upon the stiffness of the spandrel beam. Making the spandrel beams deeper and the perimeter columns more closely spaced mitigates the "shear lag" phenomenon. Placing the long sides of the rectangular columns' cross-sections along the building facade also contributes positively to the stiffness of the spandrel beams.

The behaviour of the framed-tube is obtained by placing the perimeter columns usually at 1.5 to 4.5 m centres. Closely spacing the perimeter columns and increasing the depth of the spandrel beams may test the height limits of the framed-tube system. For example, in the 110-storey, 415/417 m high World Trade Center Twin Towers (New York, 1972) (Figure 3.55), the perimeter columns were spaced at 1.02 m centres with 0.66 m in clear span (Chapter 4).

Closely spaced perimeter columns can obstruct the panoramic exterior view from inside the building and, at the ground floor, inhibit the creation of inviting public spaces with wide entrances such as lobbies and shopping centres. As a solution, with the aim of preventing the difficulties of access experienced when passing through these spaces at the building entrance, deep transfer arches or beams can be used, as in the 20-storey, 84 m high IBM Building (Seattle, 1964) (Figure 3.56a); and the 42-storey, 183 m high U.S. Bank Center (formerly First Wisconsin Center) (Milwaukee, 1973) (Figure 3.56b); or branching columns can be used, as in the 110-storey, 415/417 m high World Trade Center Twin Towers (New York, 1972) (Figure 3.56c). Below the transfer levels formed by transfer beams and branching columns, closely spaced columns are replaced with widely spaced columns.

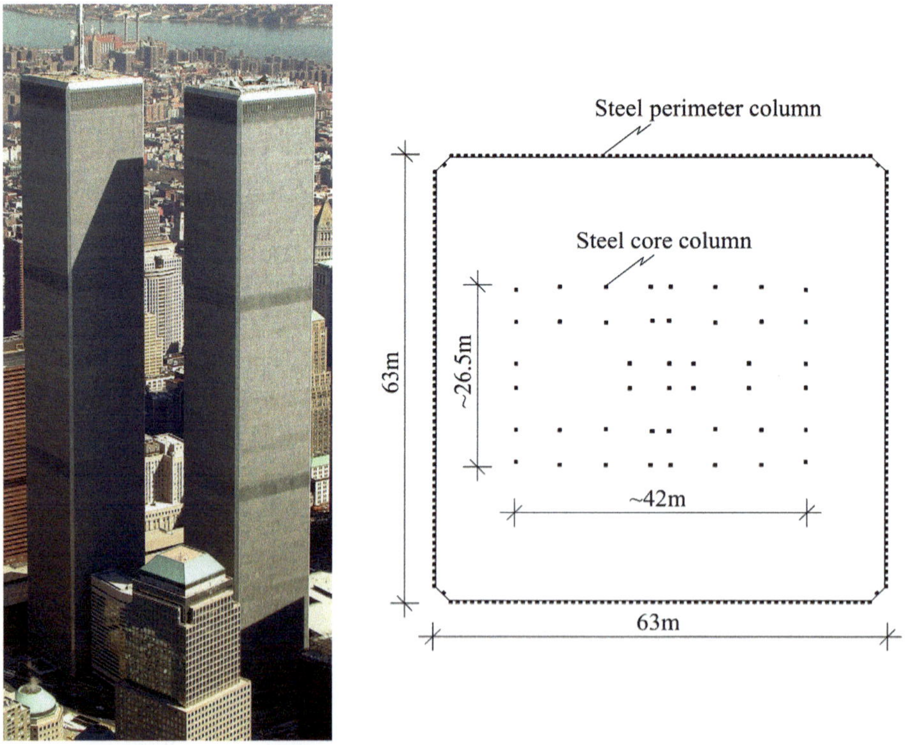

FIGURE 3.55 World Trade Center Twin Towers, New York, USA, 1972

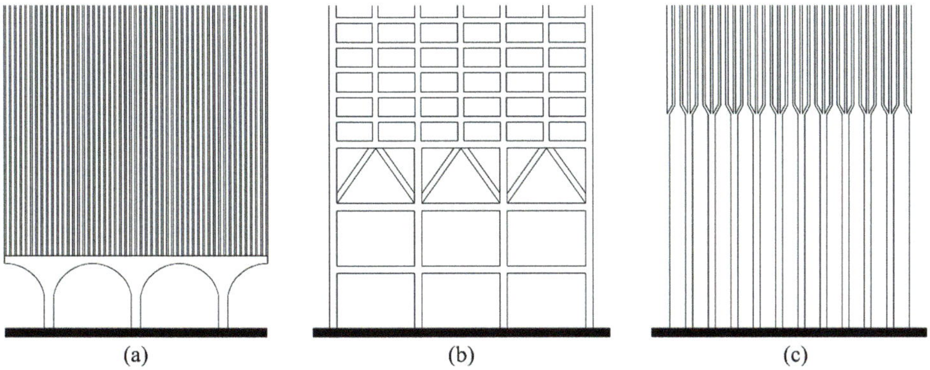

FIGURE 3.56 Configurations of the ground floor in the framed-tube system

The innovator of the idea of the framed-tube, the creative structural engineer Fazlur Rahman Khan, used it for the first time in the 43-storey, 120 m high The Plaza on Dewitt (Chicago, 1966) (Figure 3.52), which has a reinforced concrete structural system.

Some examples of tall buildings using the framed-tube system with steel structural material include:

- the 110-storey, 415/417 m high World Trade Center Twin Towers (New York, 1972) (Figure 3.55)

and with reinforced concrete structural material include:

- the 33-storey, 144 m high Torre Agbar (Barcelona, 2004) (Figure 3.57)
- the 63-storey, 223 m high Olympia Centre (Chicago, 1986) (Figure 3.58)
- the 41-storey, 167 m high First Canadian Centre (Calgary, 1982) (Figure 3.59).

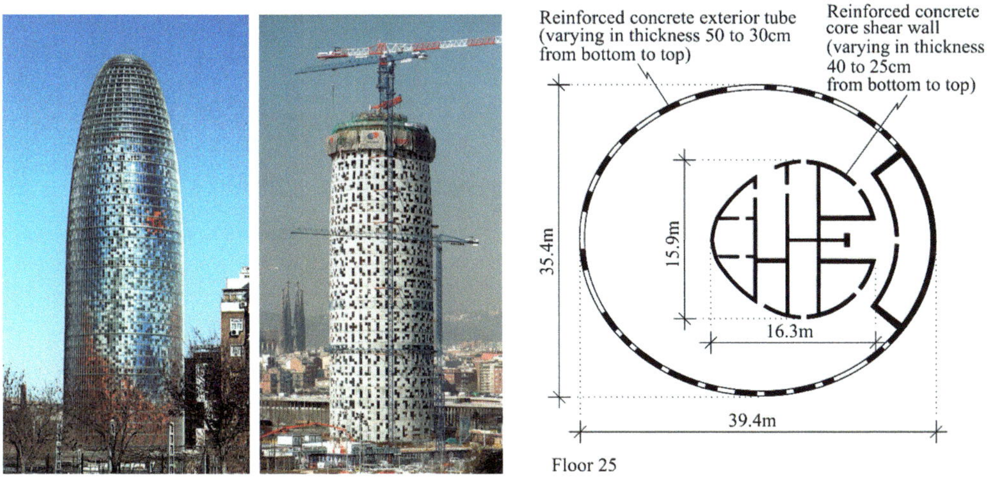

FIGURE 3.57 Torre Agbar, Barcelona, Spain, 2004
(photo on left courtesy of Niels Jakob Darger and on right courtesy of PERI Gmbh)

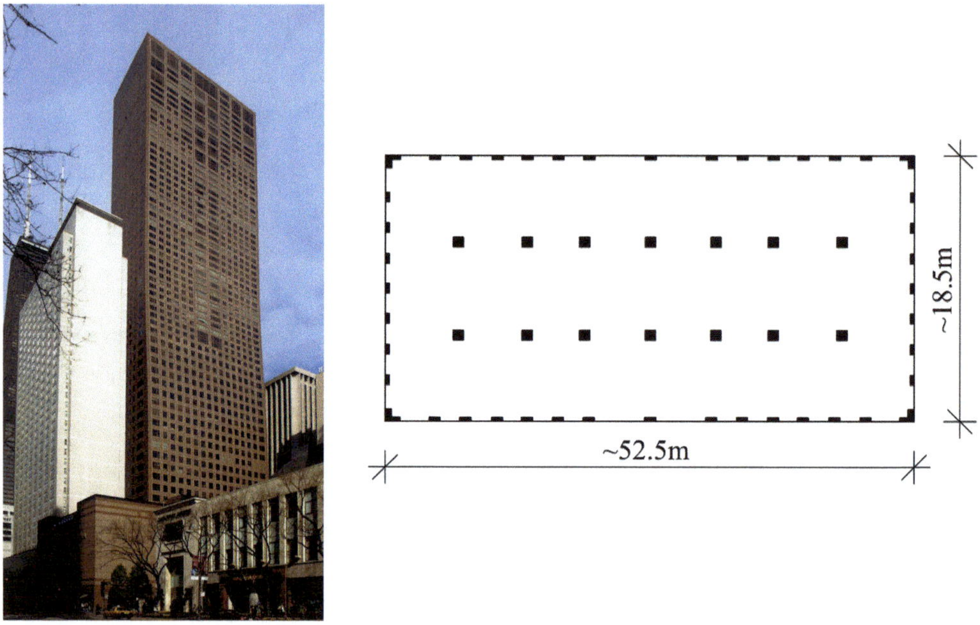

FIGURE 3.58 Olympia Centre, Chicago, USA, 1986
(photo courtesy of Marshall Gerometta/CTBUH)

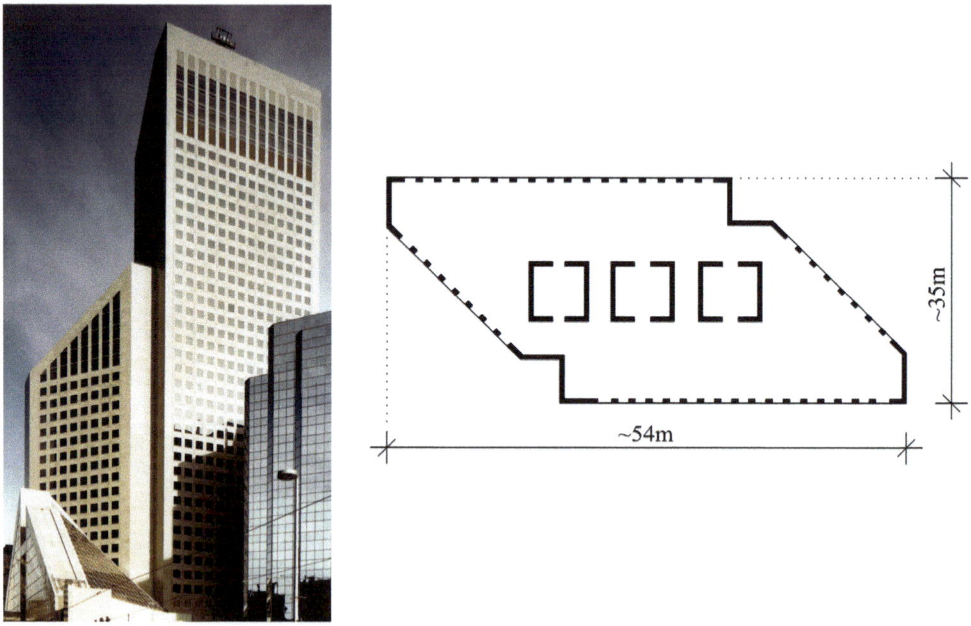

FIGURE 3.59 First Canadian Centre, Calgary, Canada, 1982
(photo courtesy of Fiona Spalding-Smith)

The diagrid-framed-tube system can be formed by using closely spaced diagonal braces instead of vertical columns (Figure 3.60). This system is more effective against lateral loads than the conventional framed-tube system. Placing the elements in a closely spaced diagrid pattern provides sufficient resistance against vertical and lateral loads. While the shear forces caused by lateral loads are met by the bending strength of the columns and beams in the framed-tube system, in the diagrid-framed-tube system they are met by the axial compressive and tensile strength of the diagonal braces. In tall buildings where lateral loads are critical, shear forces are met by axial deformation of the diagonal braces instead of bending deformation of the beams and columns, which significantly increases the efficiency of the structural system.

Some examples of tall buildings using the diagrid-framed-tube system with steel structural material include:

- the 41-storey, 180 m high 30 St Mary Axe (London, 2004) (Figure 3.61)

with reinforced concrete structural material include:

- the 40-storey, 118 m high COR Building (Miami, project pending) (Figure 3.62)
- the 22-storey, 106 m high O-14 (Dubai, 2010) (Figure 3.63)

and with composite structural material include:

- the 103-storey, 439 m high Guangzhou International Finance Center (Guangzhou, 2010) (Figure 3.64) (Ali and Moon, 2007)
- the 46-storey, 182 m Hearst Magazine Tower (New York, 2006) (Figure 3.65).

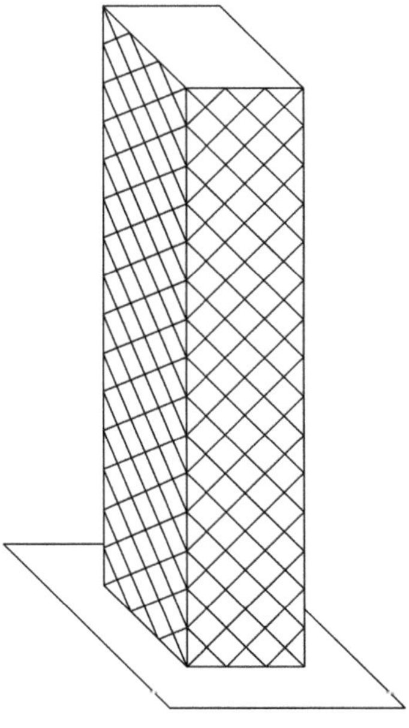

FIGURE 3.60 Diagrid-framed-tube system

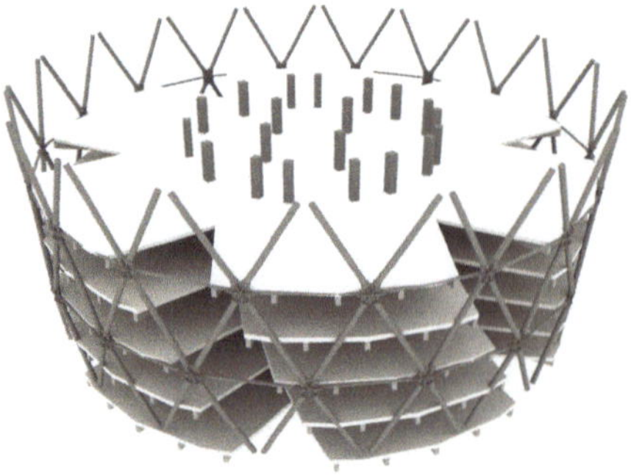

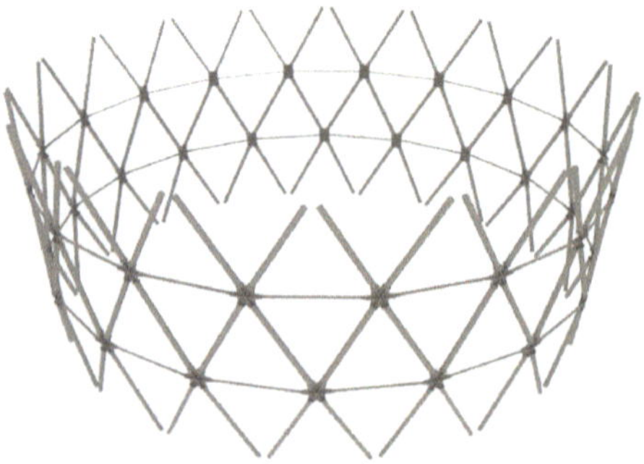

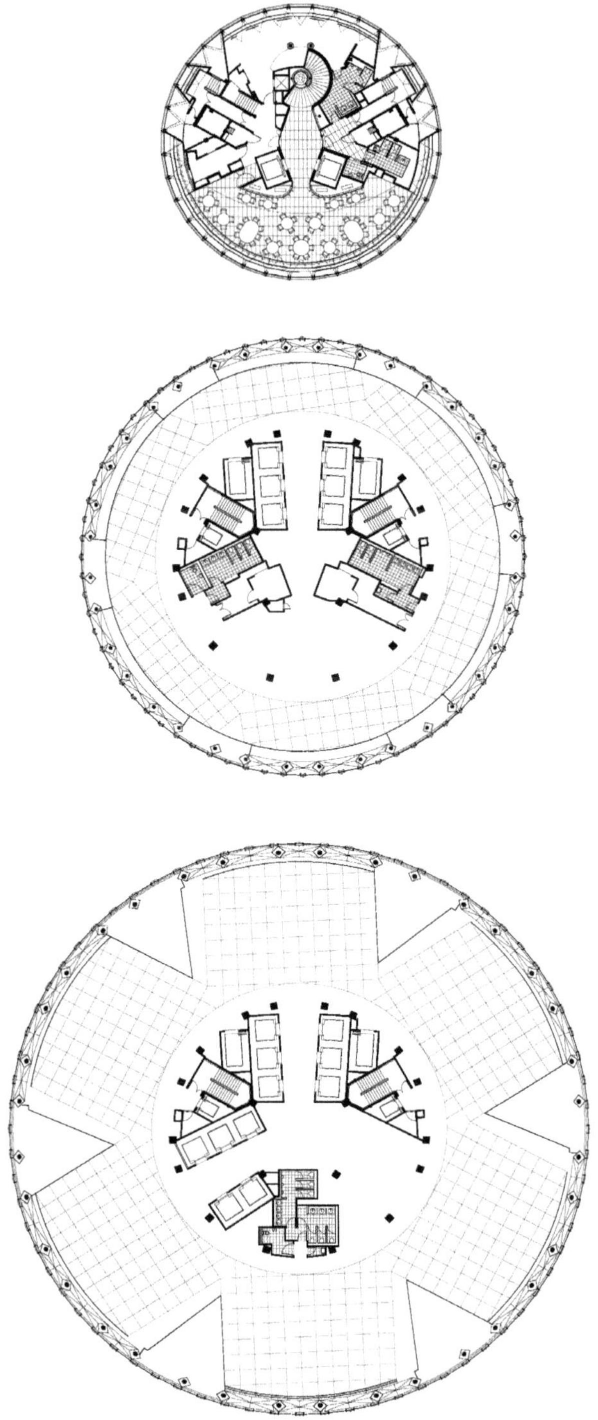

FIGURE 3.61 30 St Mary Axe, London, UK, 2004
(photos courtesy of Nigel Young/Foster + Partners; drawings on previous page courtesy of Abbas Riazibeidokhti; and drawings on this page courtesy of Foster + Partners)

FIGURE 3.62 COR Building, Miami, USA, project pending
(credit for Images: DBox Inc.)

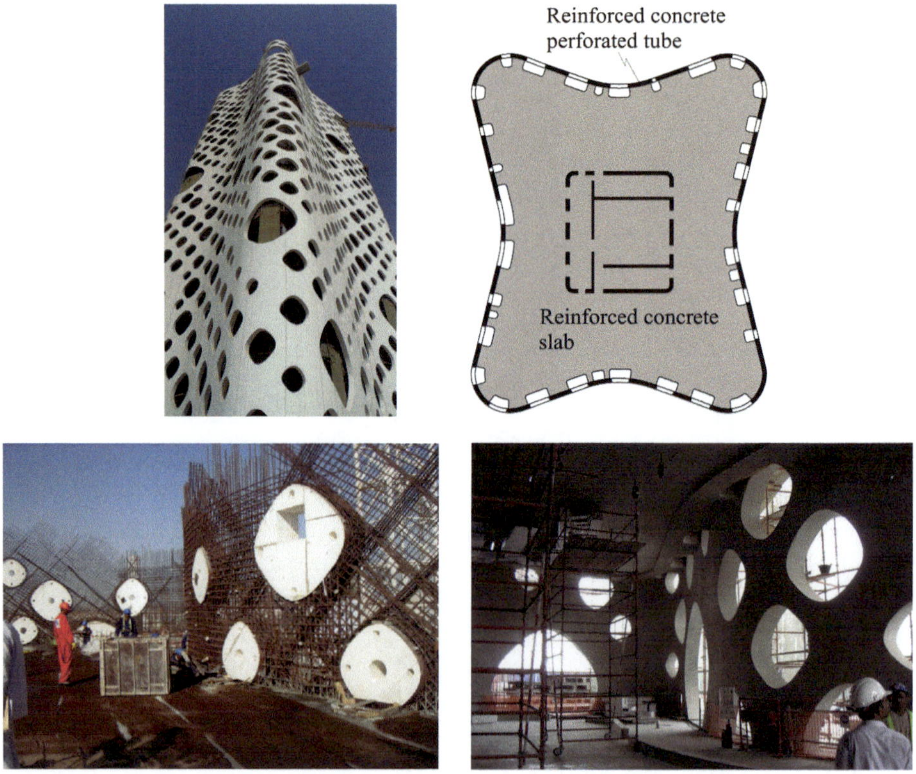

FIGURE 3.63 O-14, Dubai, U.A.E, 2010
(credit for photos: Reiser +Umemoto, RUR Architecture, PC)

FIGURE 3.64 Guangzhou International Finance Center, Guangzhou, China, 2010

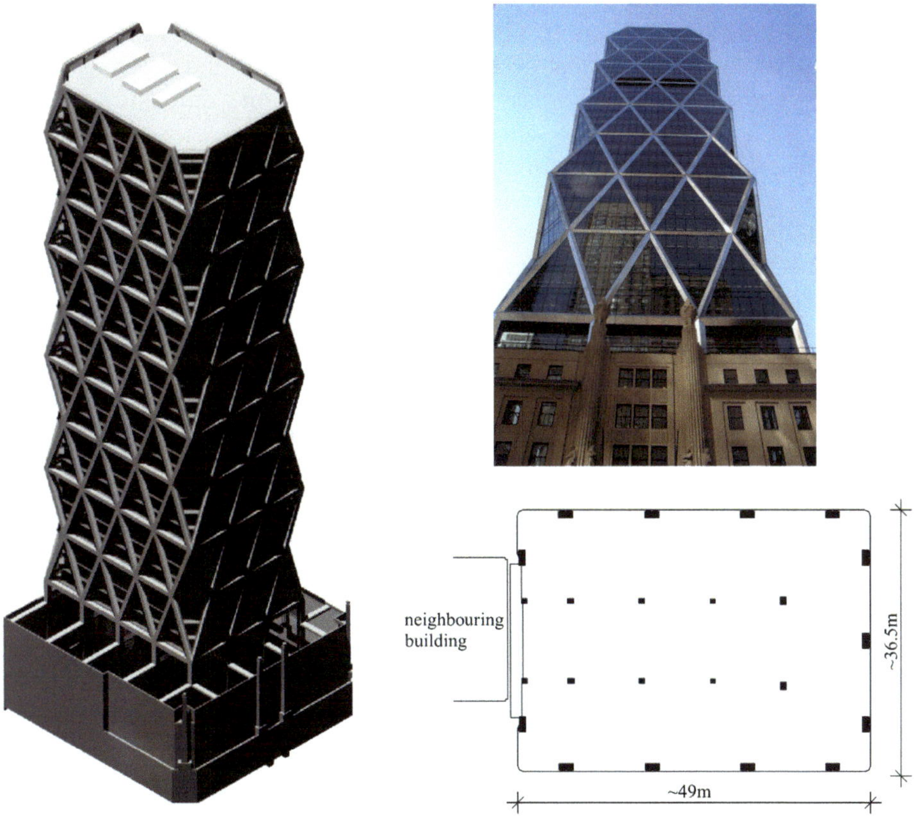

FIGURE 3.65 Hearst Magazine Tower, New York, USA, 2006
(photo on right courtesy of Antony Wood/CTBUH and drawing on left courtesy of Ozgur Ozturk)

3.10.2 Trussed-tube systems

In the framed-tube system, closely spaced perimeter columns can obstruct the panoramic exterior view from inside the building. In order to increase the spacing between the columns without inhibiting the tubular behaviour, connecting the perimeter columns with exterior multi-storey braces led to the development of the trussed-tube (braced-tube) system (Figure 3.68). Trussed-tube system can be described as the improvement of the framed-tube system, and it was likewise innovated by Fazlur Rahman Khan.

Adding braces to the exterior of the framed-tube system makes it approach very closely pure tubular cantilever behaviour by increasing the structural stiffness, effectiveness, and reduces the negative effect of the "shear lag" caused by the flexibility of the spandrel beams. Compared with the framed-tube system, the trussed-tube system gives scope for increasing the height of the structure with wider spacing between columns. As in the case of the 59-storey, 279 m high Citigroup Center (New York, 1977) (Figure 3.66) and the 100-storey, 344 m high John Hancock Center (Chicago, 1969) (Figure 3.67), maximum column spacing is 11.5 m and 13.3 m centres respectively.

Fazlur Rahman Khan emphasised that the exterior braces, which made it possible to have wide spaces between the columns, would behave like inclined columns, and moreover they transferred load to or from the columns by allowing redistribution of the stresses resulting with almost evenly load distribution in the columns. According to Khan, this system would increase the structural system's efficiency and that this would allow the construction of supertall buildings.

In buildings with steel or composite trussed-tube systems, multi-storey braces (diagonal or X-braces) are used on the facade of the building (Figure 3.68a). In the case of buildings with reinforced concrete trussed-tube systems, spaces between the columns are filled with reinforced concrete shear walls to form multi-storey diagonal or X-brace pattern on the exterior of the building (Figure 3.68b).

Fazlur Rahman Khan used the trussed-tube system for the first time in the 100-storey, 344 m high John Hancock Center (Chicago, 1969), with a steel structural system (Figure 3.67). The 50-storey, 174 m high 780 Third Avenue Building (New York, 1983) (Figure 3.69) was the first reinforced concrete building in which a trussed-tube system was used.

Some examples of tall buildings using the trussed-tube system with reinforced concrete structural material include:

- the 58-storey, 174 m high Onterie Center (Chicago, 1986) (Figure 3.70)

and with composite structural material include:

- the 72-storey, 367 m high Bank of China Tower (Hong Kong, 1990) (Figure 3.71) (Colaco, 2005; Kijewski-Correa, 2002)
- the 59-storey, 279 m high Citigroup Center (New York, 1977) (Figure 3.66)
- the 49-storey, 234 m high CCTV Headquarters (Beijing, 2011) (Figure 3.72).

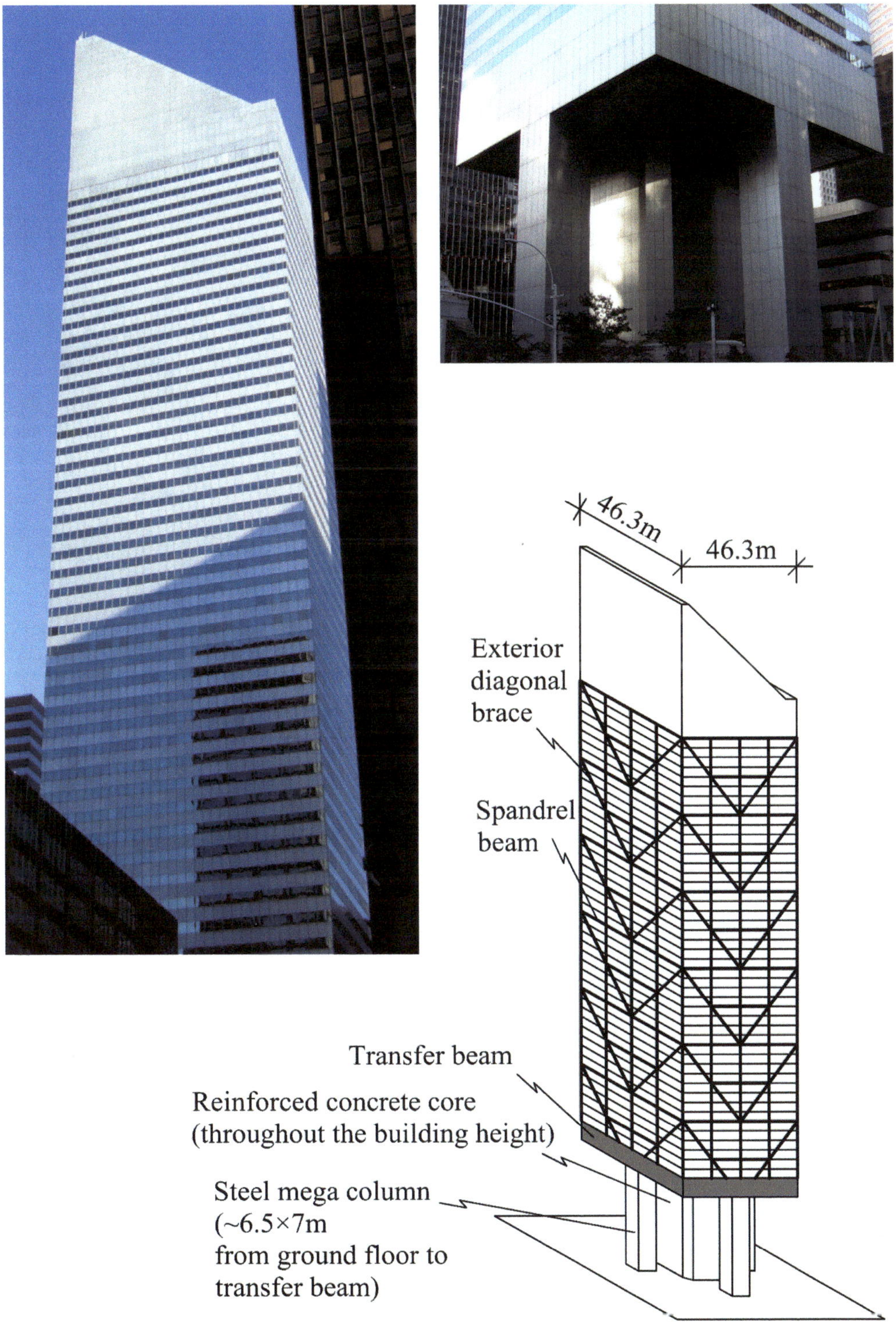

FIGURE 3.66 Citigroup Center, New York, USA, 1977

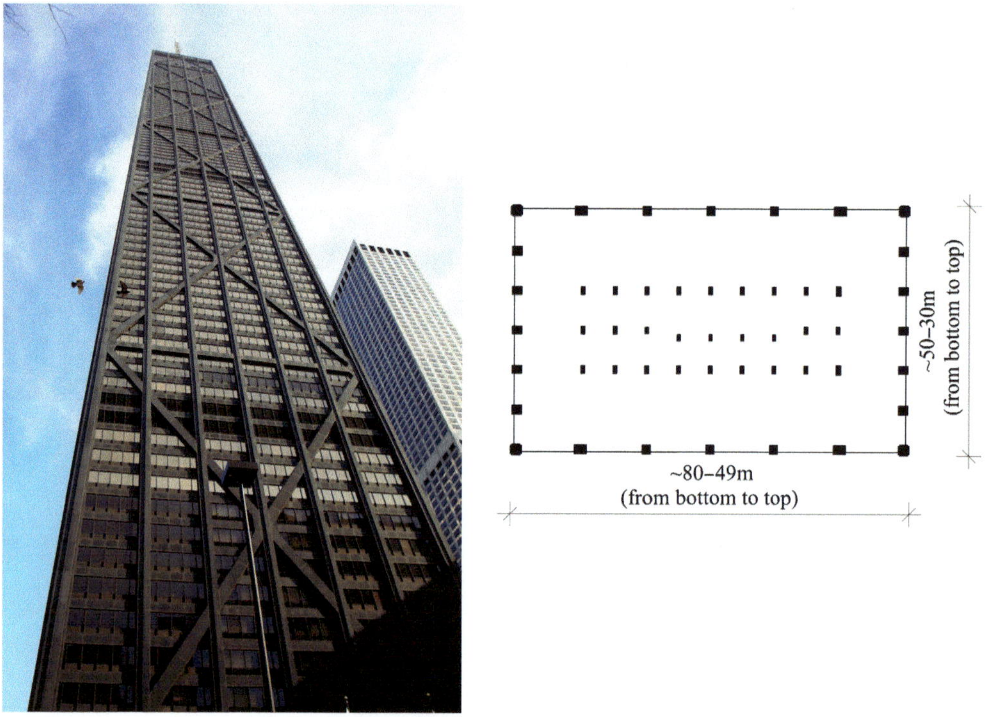

FIGURE 3.67 John Hancock Center, Chicago, USA, 1969
(photo courtesy of Marshall Gerometta / CTBUH)

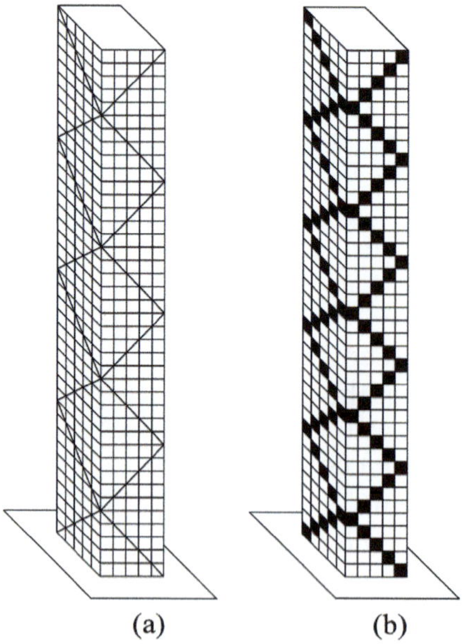

FIGURE 3.68 Trussed-tube system: (a) Steel or composite, (b) Reinforced concrete

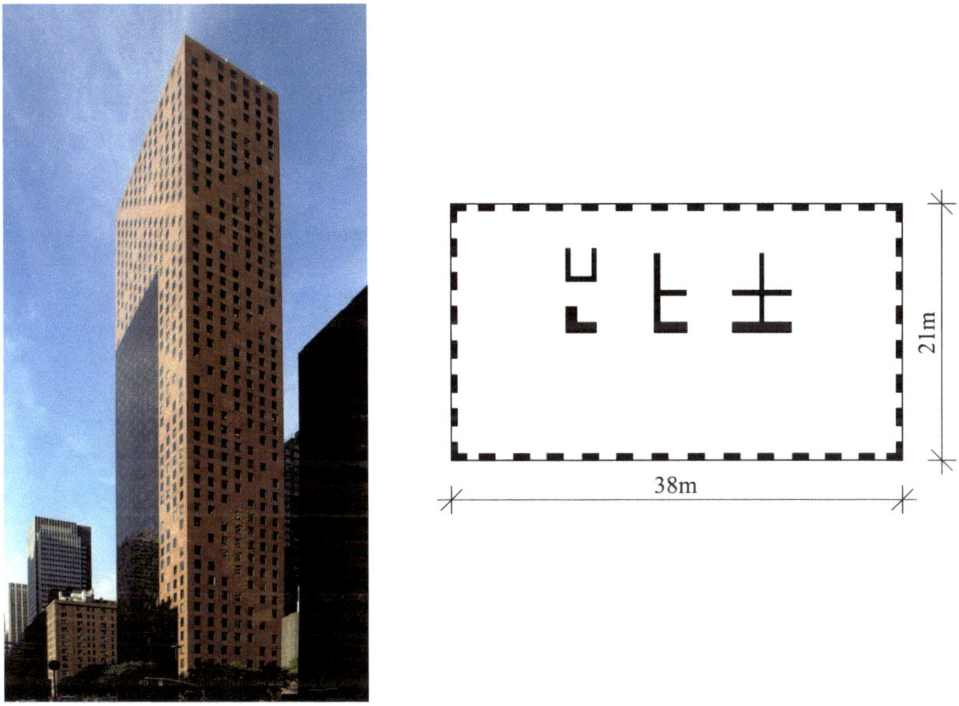

FIGURE 3.69 780 Third Avenue Building, New York, USA, 1985
(photo courtesy of Marshall Gerometta/CTBUH)

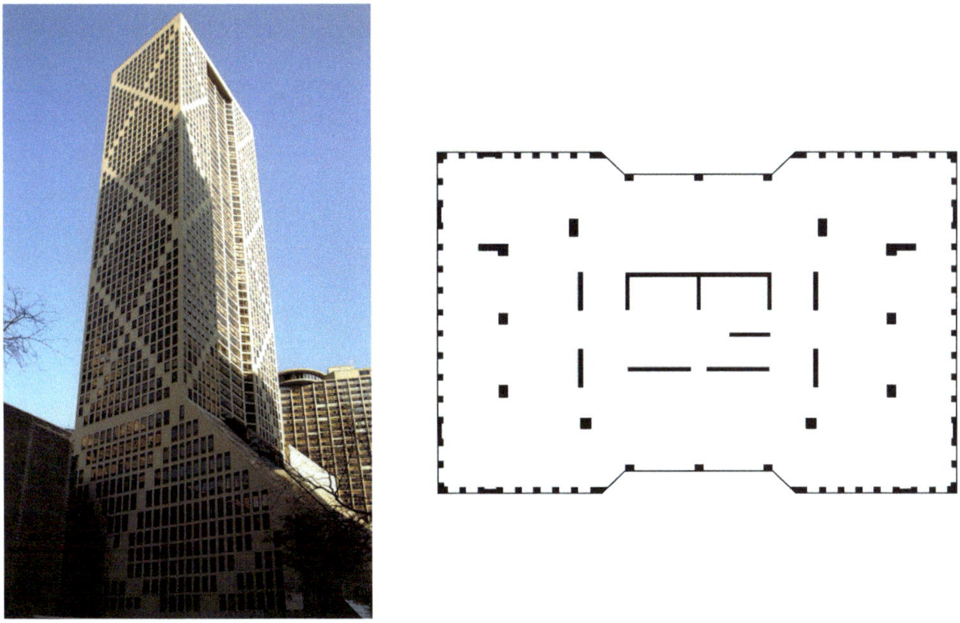

FIGURE 3.70 Onterie Center, Chicago, USA, 1986
(photo courtesy of Marshall Gerometta/CTBUH)

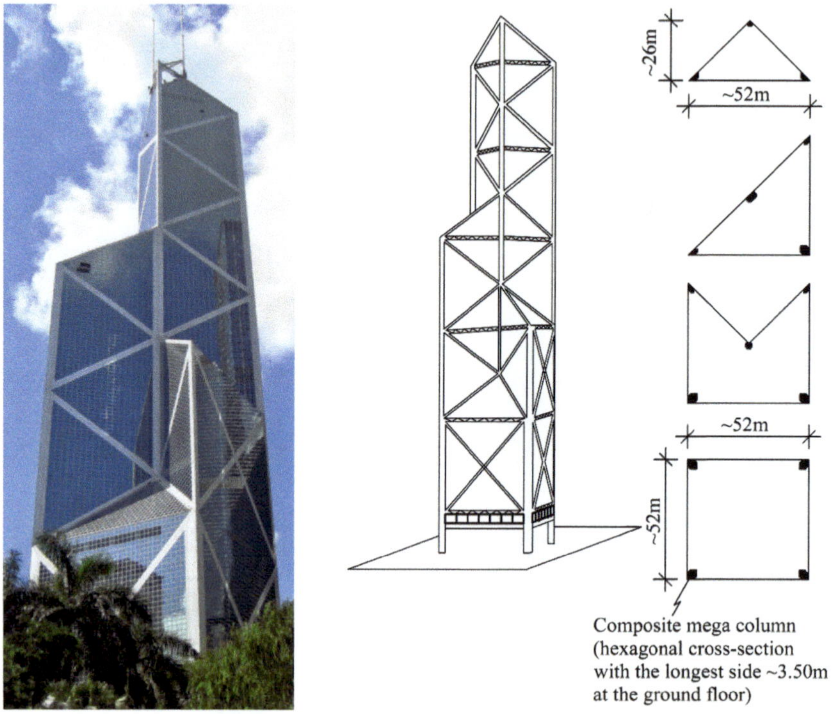

FIGURE 3.71 Bank of China Tower, Hong Kong, China, 1990

FIGURE 3.72 CCTV Headquarters, Beijing, China, 2011
(courtesy of M.Bunyamin Bilir)

Ali and Armstrong (1995), and Ali and Moon (2007) define the system used in one of these examples, the Bank of China Tower, as a "space truss system, being a development of the trussed-tube system". According to the authors, because of its four mega columns at the corners and the mega braces which connect them, it is possible to classify the Bank of China Tower's structural system not only as a trussed-tube system, but also as a mega column, a mega frame, or a space truss system. In appearance although it resembles a structure with distinct tubes terminating at several layers, the Bank of China's structural system is not a bundled-tube system.

3.10.3 Bundled-tube systems

Bundled-tube systems are a combination of more than one tube (framed-tube and/or trussed-tube) acting together as a single tube (Figure 3.73). Like the framed-tube and trussed-tube systems, the bundled-tube system was also innovated by the structural engineer Fazlur Rahman Khan. Among the advantages of the bundled-tube system are: the securing of architectural freedom thanks to the ability to create tubes of different heights in the system; the attainment of higher building heights and wider column spaces than in framed-tube systems; and the ability to control the aspect ratio.

In the bundled-tube system, setbacks with floor plans of different shapes and dimensions are obtained by ending tubes at the desired levels. Single tubes in the system can be arranged together in different shapes such as rectangles and triangles, and thus different forms can be created.

As the heights of buildings increase, in general their aspect ratios also increase. The increase in the aspect ratio increases the slenderness and flexibility of the building, and thus its lateral drift. In order to keep control of the aspect ratio, it is necessary to increase the cross-sectional dimensions of the base, which affects the denominator in this ratio. In bundled-tube systems consisting of two or more tubes, the tubes can rise to different levels of the building height (Figure 3.73). Thus, in bundled-tube systems, the increase in the cross-sectional dimensions at the ground floor in order to control the slenderness of the building makes it possible to reduce the cross-sectional dimensions by different amounts throughout the height of the building.

In bundled-tube systems formed from framed-tubes and/or trussed-tubes, greater building heights and wider column spaces are obtained than in framed-tube systems. For example, in the Willis Tower (Chicago, 1974) (Figure 3.73), which has 9 framed-tubes, the spaces between the columns are much greater than the column spaces in a framed-tube building of the same height. While the 110-storey, 415/417 m high World Trade Center Twin Towers had perimeter columns spaced at 1.02 m centres, the 108-storey, 442 m high Willis Tower has perimeter columns spaced at 4.6 m centres.

Some examples of tall buildings using the bundled-tube system with steel structural material include:

- the 108-storey, 442 m high Willis Tower (Chicago, 1974) (Figure 3.73) and with reinforced concrete structural material include:

- the 57-storey, 205 m high One Magnificent Mile (Chicago, 1983) (Figure 3.74)

and with composite structural material include:

- the 55-storey, 233 m high Wachovia Financial Center (Miami, 1983).

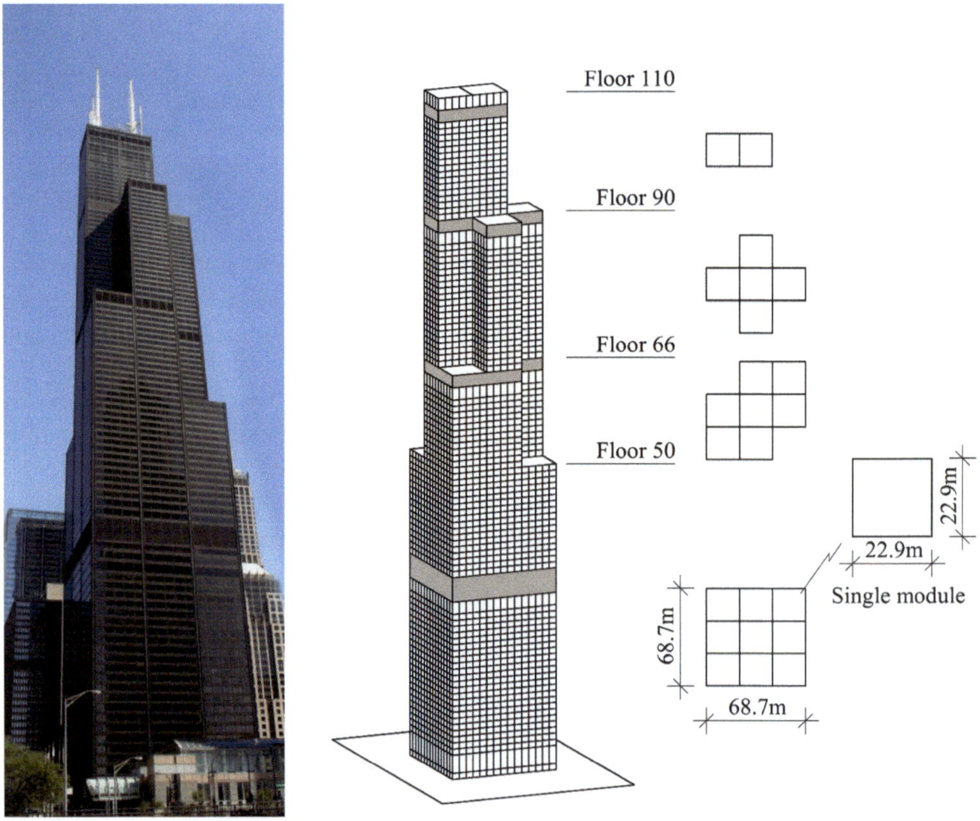

FIGURE 3.73 Willis Tower, Chicago, USA, 1974
(photo courtesy of Antony Wood/CTBUH)

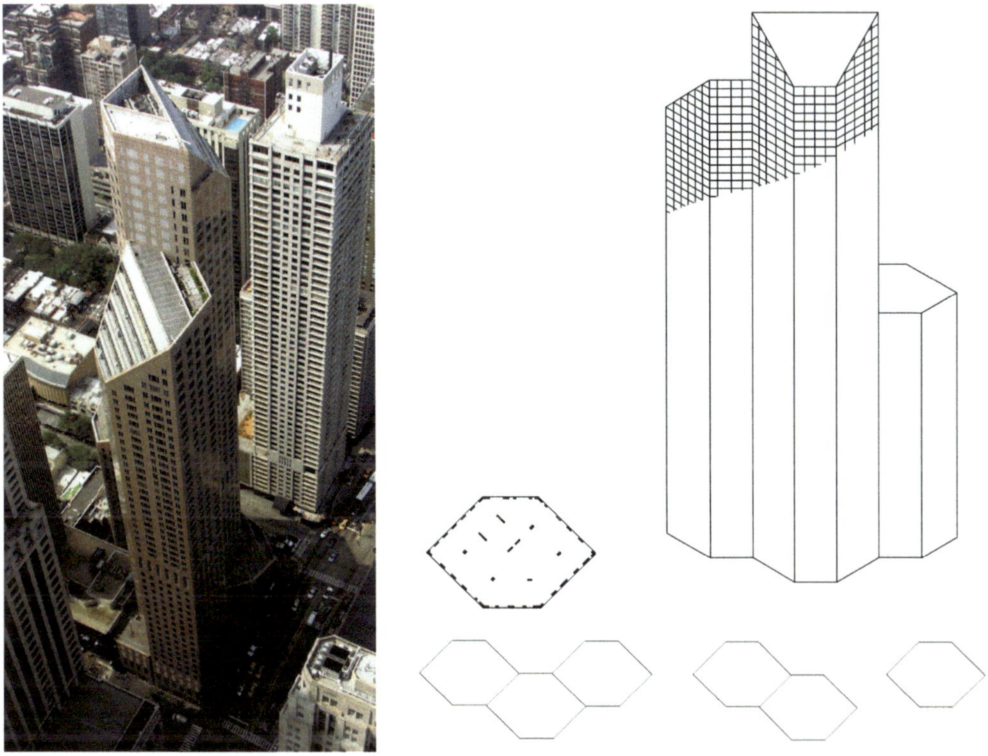

FIGURE 3.74 One Magnificent Mile, Chicago, USA, 1983 (Kim and Elnimeiri, 2004) *(photo courtesy of Marshall Gerometta/CTBUH)*

4

TALL BUILDING CASE STUDIES

Home Insurance Building

OFFICIAL NAME: Home Insurance Building
LOCATION: Chicago, USA
BUILDING FUNCTION: Office
ARCHITECTURAL HEIGHT: 55 m
NUMBER OF STOREYS: 10+2 (added in 1890)
STATUS: Demolished to make way for a new building (1931)
COMPLETION: 1885
ARCHITECT: William Le Baron Jenney
STRUCTURAL ENGINEER: William Le Baron Jenney
STRUCTURAL SYSTEM: Rigid frame system/steel

Home Insurance Building: architectural and structural information

The 12-storey, 55 m high, Home Insurance Building in Chicago (USA) was designed by engineer William Le Baron Jenney. It is a steel building with a rigid frame system. Known as the "father of skyscrapers", the Home Insurance Building is regarded as the first skyscraper.

The Home Insurance Building opened a new era in tall building construction and took the title of "first skyscraper", owing to Le Baron Jenney's idea of using iron and steel elements for the structure of a tall building instead of thick stone walls and as a development of this, his discovery of the frame structural system consisting of a skeleton of horizontal beams and vertical columns. The frame structural system used in the Home Insurance Building became a model for subsequent tall building designs.

Although it did not break the height record, the Home Insurance Building is unique in being the first building designed with a frame structural system consisting of iron and steel elements, instead of masonry walls, as structural support. Compared with a building with load-bearing masonry walls and the same number of storeys, the metal skeleton was three times lighter and the exterior walls were designed only to protect the building from adverse weather conditions and not to provide structural support. It was recognised as having many more and larger window openings in its facade than other buildings of its era.

With the commencement of the use of non-structural perimeter and partition elements in tall buildings instead of thick masonry load-bearing walls, thanks to the appearance of the Home Insurance Building and the frame system, the architectural and financial value of the spaces within buildings was increased and more usable space and natural light was obtained in the interior. Thus, just as this development made the construction of tall buildings much more practical, it also increased their utility.

Bibliography

Bennett, D., *Skyscrapers: Form & Function*, Simon & Schuster Ltd., New York, 1995.

Harbert, L., Home Insurance Building – The First Skyscraper, *Journal of American Society of Civil Engineers* (ASCE), Vol. 43, No. 2, pp. 1–2, 2002.

Home Insurance Building, http://www.ctbuh.org, accessed May 2012.

Home Insurance Building, http://www.emporis.com, accessed May 2012.

Empire State Building

OFFICIAL NAME: Empire State Building
LOCATION: New York, USA
BUILDING FUNCTION: Office
ARHITECTURAL HEIGHT: 381 m
NUMBER OF STOREYS: 102
STATUS: Completed
COMPLETION: 1931
ARCHITECT: Shreve Lamb & Harmon Associates
STRUCTURAL ENGINEER: H.G. Balcom & Associates; Post and McCord; Strong & Jones Engineers
STRUCTURAL SYSTEM: Shear trussed frame system/steel

(courtesy of Antony Wood/CTBUH)

Empire State Building: architectural and structural information

The 102-storey, 381 m high Empire State Building in New York (USA) was designed by Shreve Lamb & Harmon Associates. It is a steel building with a shear trussed frame system. The Empire State Building gained the title of "the world's tallest building" in 1931. Currently, it has held the title of "the world's tallest building" for the longest period (41 years, 1931–1972).

Legendary as a national historical monument symbolising American courage and skill, the Empire State Building, with the style of its facade and its long-held title of "the world's tallest building", shaped the city's skyline as the spirit of the Art Deco period and New York's international architectural icon.

The concrete-encased braced frames (shear trusses), which give the Empire State Building's structural system its basic character (Figure 4.1), were designed to resist the entire vertical and lateral loads with the help of a stone cladding on the facade. Although it is not included in the calculations for the structural analysis, the concrete encasement makes a significant contribution to the shear-trussed frame's strength against wind induced lateral loads.

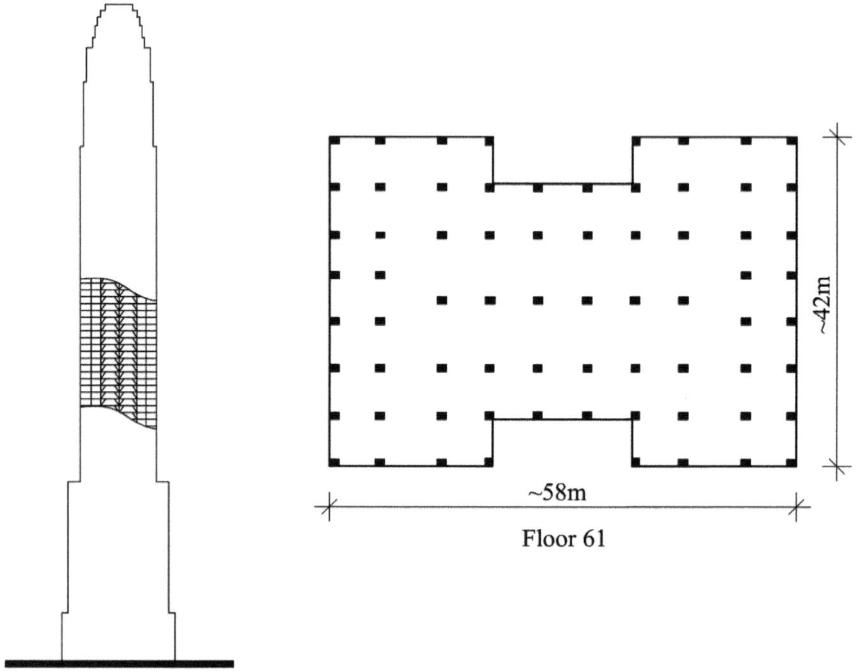

FIGURE 4.1 Empire State Building plan and section

Bibliography

Empire State Building, http://www.ctbuh.org, accessed May 2012.

Empire State Building, http://www.emporis.com, accessed May 2012.

Taranath, B., *Steel, Concrete & Composite Design of Tall Buildings*, New York: McGraw-Hill Book Company, 1998.

Taranath, B., *Wind and Earthquake Resistant Buildings: Structural Analysis and Design*, A Series of Reference Books and Textbooks (Editor: Michael D. Meyer), Department of Civil and Environmental Engineering, Georgia Institute of Technology, Atlanta, Georgia, 2005.

Zaknic, I., Smith, M. and Rice, D., *100 of The World's Tallest Buildings*, Council on Tall Buildings and Urban Habitat Committee, New York: McGraw-Hill Book Company, 1998.

Bahrain World Trade Center (BWTC)

OFFICIAL NAME: Bahrain World Trade Center
LOCATION: Manama, Kingdom of Bahrain
BUILDING FUNCTION: Mixed-use (office, commercial)
ARCHITECTURAL HEIGHT: 240 m
NUMBER OF STOREYS: 45
STATUS: Completed
COMPLETION: 2008
ARCHITECT: WS Atkins & Partners
STRUCTURAL ENGINEER: WS Atkins & Partners
STRUCTURAL SYSTEM: Shear walled frame system/reinforced concrete

(credit for photo: WS Atkins & Partners, Middle East)

Bahrain World Trade Center: architectural and structural information

The 45-storey, 240 m high Bahrain World Trade Center in Bahrain (Manama) was designed by WS Atkins & Partners. It is a reinforced concrete building with a shear walled frame system. The BWTC was recognised as the winner of the award: "Best Tall Building 2008, Middle East and Africa" by CTBUH.

The BWTC is the first in the world to integrate large-scale wind turbines on to a building. The main design idea of the building was inspired by the aerodynamics of sails and traditional Arabian wind towers. The BWTC draws attention not only to sail-shaped twin towers but also to innovative and pioneer approaches towards sustainability. The building makes new attempts with regard to sustainable architecture by converting wind energy to electrical energy to reduce the total energy

FIGURE 4.2 Site of BWTC
(credit for photo: WS Atkins & Partners, Middle East)

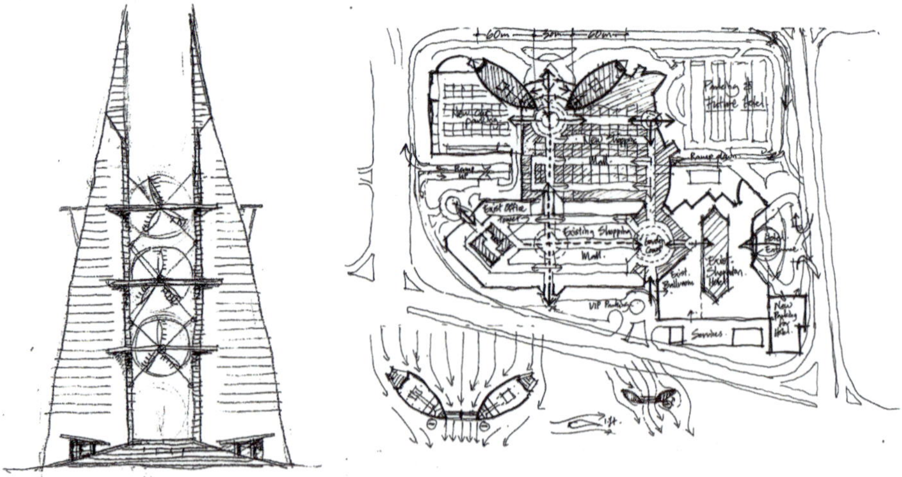

FIGURE 4.3 Initial sketches by Shaun Killa
(credit for drawings: WS Atkins & Partners, Middle East)

consumption of the building. The premium for including the wind turbines cost less than 3 per cent of the project value but on the other hand 11–15 per cent of the electrical energy consumption of the building is expected to be generated. Besides the wind turbine renewables, the building includes some other sustainable features, such as buffer spaces between the external environment and air conditioned spaces, shading devices for sun control, energy efficient lighting system with zonal control and dual drainage systems for water recycling etc.

Located in the downtown central business of Manama, BWTC has an impressive site overlooking the Arabian Gulf. At the beginning of the site analysis, Atkins' principal architect and keen sailor Shaun Killa realised the potential of the prevailing onshore wind having direction almost perpendicular to the site (Figure 4.2). After the analysis of the wind, the architect has decided to convert this condition into an advantage by emphasising a well-designed sustainable solution.

According to the design inspiration idea, for harnessing the prevailing north-westerly onshore breeze from the Gulf and using its energy, the architect formed the twin towers and integrated the horizontal-axis wind turbines with the help of the bridges between them (Figure 4.3).

The fact that the wind turbines can generate energy depends on the consistent wind energy with the capability of the horizontal wind turbine blades adjusting themselves to the prevailing wind direction. However, in the BWTC, the blades of the turbines could not adjust themselves to changing wind directions due to their standstill positions on the bridges. To cope with the problem, the architect focused on the shape and the position of the towers directly and decided to design the building in such a way to obtain optimum onshore wind flow among the towers. In this context, the design process is based on an "aerodynamic architectural design approach" (Section 6.1) by using wind tunnel testing together with Computational Fluid Dynamics (CFD) Modelling.

The elliptical plan forms funnelling the onshore breeze augment the wind flow and thus accelerate the wind velocity (up to 30 per cent amplifying the wind speed) and create a negative pressure behind the towers. The tapering shape of the towers towards the top reduces the effect of the augmentation of the wind flow. This effect, together with the velocity profile of the wind being lowest at the ground level, allows all the three wind turbines to rotate almost at the same speed and thus generating the same energy. Wind tunnel tests show that when the towers are subjected to onshore wind not perpendicular but with oblique angles of 45° in direction, the centre of the wind stream remains nearly perpendicular to the turbines.

Between the BWTC towers, to support wind turbines, three steel V-shape bridges (173° to avoid blade strike during extreme conditions) spanning 31.7 m in length are connected to the towers at different levels. They allow the towers to move 0.5 m towards each other. Three 29 m diameter horizontal axis wind turbines are located at bridge levels 60 m, 96 m and 132 m respectively above the ground (Figure 4.4).

In plan, on one side of the towers, there is a service area placed close to the wind turbines to prevent undesirable noise and vibration in offices and the shopping mall area. On the other side of the towers, there are two different types of columns: one is circular in cross-section and angular in both directions and the other is rectangular in cross-section; both are inclined almost 15° (Figure 4.5–4.6).

FIGURE 4.4 BWTC horizontal axis wind turbines
(credit for photo: WS Atkins & Partners, Middle East)

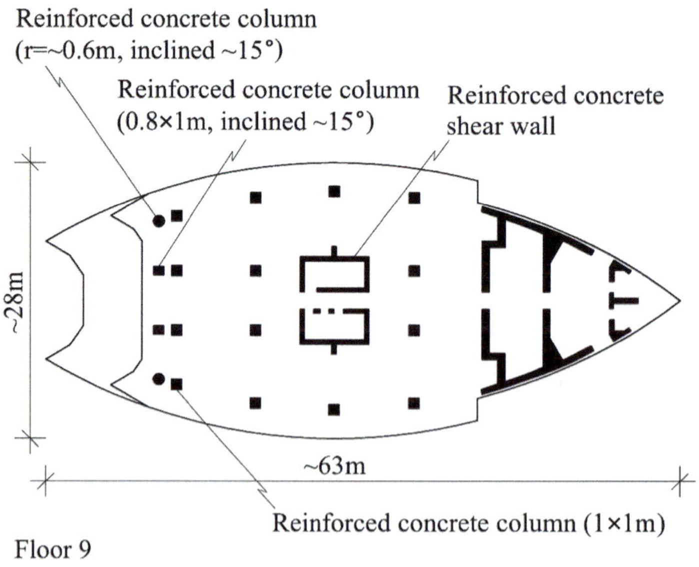

FIGURE 4.5 BWTC plan

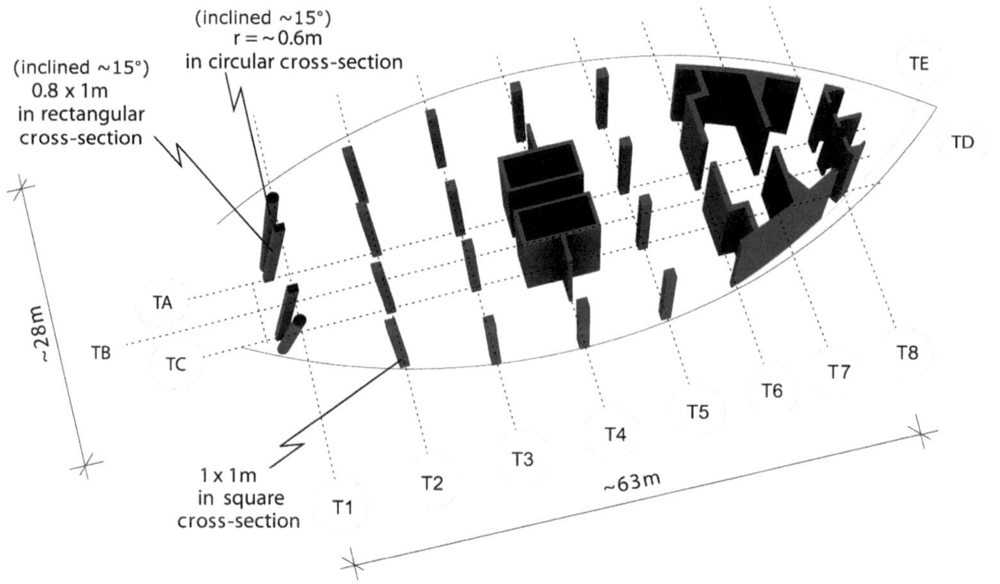

FIGURE 4.6 BWTC structural axonometric

Bibliography

Bahrain World Trade Center, http://bahrainwtc.com, accessed May 2012.

Bahrain World Trade Center, http://www.ctbuh.org, accessed May 2012.

Bahrain World Trade Center, http://www.emporis.com, accessed May 2012.

Bellini, O.E. and Daglio, L., *Bahrain World Trade Center*, Building Arabia, White Star Publishers, pp. 36–40, Vercelli, 2009.

BWTC Scores World First, *Gulf Construction Magazine*, pp. 71–76, May 2008.

Smith, R.F. and Killa, S., Bahrain World Trade Center (BWTC): The First Large-Scale Integration of Wind Turbines in a Building, *The Structural Design of Tall and Special Buildings*, Vol.16, pp. 429–439, 2007.

WS Atkins & Partners, *Bahrain World Trade Center: The World's First Large Scale Integration of Wind Turbines on a Building, CTBUH Best Tall Building Awards*, pp. 1–21, WS Atkins & Partners Publication, 2008.

Pirelli Building

OFFICIAL NAME: Pirelli Building
LOCATION: Milan, Italy
BUILDING FUNCTION: Office
ARCHITECTURAL HEIGHT: 127 m
NUMBER OF STOREYS: 32
STATUS: Completed
COMPLETION: 1958
ARCHITECT: Gio Ponti
STRUCTURAL ENGINEER: Pier Luigi Nervi
STRUCTURAL SYSTEM: Shear walled frame system/reinforced concrete

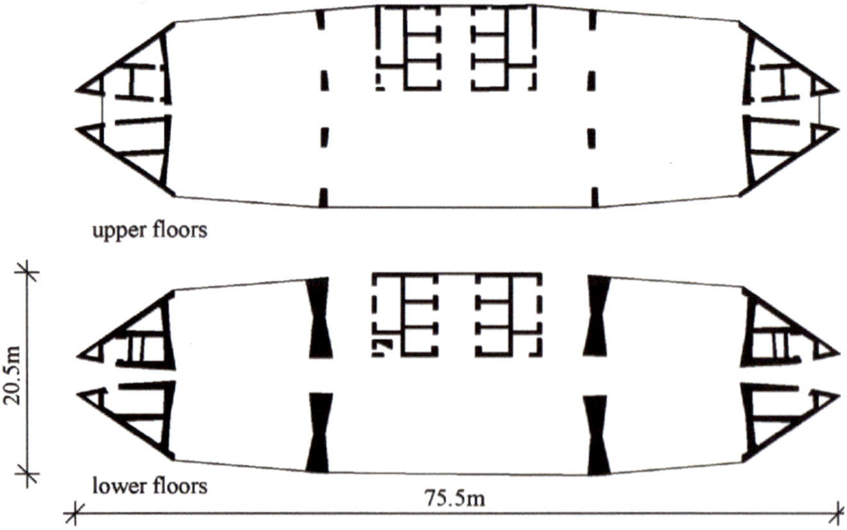

FIGURE 4.7 Pirelli Building plans

Pirelli Building: architectural and structural information

The 32-storey, 127 m high Pirelli Building in Milan (Italy) was designed by Gio Ponti. It is a reinforced concrete building with a shear walled frame system. The Pirelli Building is the first building utilising the interactive system of frames and shear walls. It was the highest building with reinforced concrete structure and gained the title of "the world's tallest reinforced concrete building" at that time in Europe. The Pirelli Building became a symbol of Milan and national economic development.

The innovation of utilising the interaction between rigid frames and shear walls to carry the lateral loads was first used by Pier Luigi Nervi in the structural design of the Pirelli Building. Architect Gio (Giovanni) Ponti and structural engineer Pier Luigi Nervi designed and realised the Pirelli Building together. The building is original and unique in terms of its architectural merits as well as structural inventiveness.

The Pirelli Building, being a different shape to a customary block form, has a tapered plan where the width of the building reduces towards the edges. The ground floor has a convex lens shaped plan with a length of 75.5 m and a maximum width of 20.5 m (Figure 4.7).

At the two sides of the building, two rigid triangular reinforced concrete core shear walls that hold the service areas were not sufficient for wind and earthquake induced lateral loads. Therefore, two reinforced concrete mega shear walls tapering toward the top were also added. The central core shear walls also help the stability of the structure. A large scale (1:15) model (about 10 m) was created and a series of tests were carried out in order to verify the effects of the action of the wind and earthquake induced lateral loads.

Gio Ponti (1891–1979) had his own philosophy of design which was expressed as architecture as a work of art should inherit the following qualities:

- formal and structural inventiveness
- essentiality
- representativeness
- expressiveness
- illusiveness
- perpetuity.

Gio Ponti insists that the value of "closed form – forma finite" is one of the constituent elements of the essentiality. Since he defined architecture as a work of art (the work of art, architecture, is permanent, perpetual; it cannot be repeatable and imitable), it should have a complete and finite form so that nothing could be added, nor subtracted from it. During the design process of the Pirelli Building, he achieved this principle by making the building ends like a knife-edge, thus creating a finished and closed form.

Gio Ponti and his collaboraters created a beautiful purist sculpture in the heart of Milano as well as a symbol of the power and prosperousness of Italy. E. Kaufman commented "A new look in the old American art of the skyscraper" (quoted in Ponti, 1956, p. 164).

Bibliography

Banham, R., *Age of the Masters*, Architectural Press Ltd., pp. 116-117, 1977.
Chiorino, M. and Chiorino, C., *Pier Luigi Nervi: Architecture as Challenge*, Milan, 2010. (Editors: Olmo Carlo, Cristiana Chiorino, Exhibition held at Palazzo Giustinian Lolin, Venice, p. 237.)
Curtis, W. J. R., *Modern Architecture Since 1900*, Phaidon Press Ltd., London, p. 478, 1996.
Domus, *La Torre Pirelli a Milano*, No. 348, pp. 1–3, 1958.
Kidder Smith, G. E., *The New Architecture of Europe*, Pelican Books, pp. 179–180, 1962.
Millais, M., *Building Structures: From Concepts to Design*, Taylor and Francis Books, 2005.
Pirelli Building, http://www.ctbuh.org, accessed May 2012.
Pirelli Building, http://www.emporis.com, accessed May 2012.
Ponti, G., Out of Philosophy of Architecture, *Architectural Record*, pp. 155–164, 1956.
Ponti, G., Espressione, dell'edificio Pirelli in costruzione a Milano, *Domus*, No. 316, pp. 1–16, 1956.
Ponti, G., Immeuble Pirelli, Milan, *L'Architecture d'Aujourd'hui*, No. 64, pp. 2–5, 1956.
Ponti, G., *A specific letter in which Gio Ponti explained his design philosophy was sent to Prof. Enis Kortan*, 27 October 1977.
Ziegler, C., Out of Ashes and Rubble: The Pirelli Tower, *Places*, College of Environmental Design, UC Berkeley, Vol. 21, Issue 1, 2009.

Aspire Tower

OFFICIAL NAME: Aspire Tower
LOCATION: Doha, Qatar
BUILDING FUNCTION: Mixed-use (hotel, office)
ARCHITECTURAL HEIGHT: 300 m
NUMBER OF STOREYS: 36
STATUS: Completed
COMPLETION: 2006
ARCHITECT: Hadi Simaan
STRUCTURAL ENGINEER: Ove & Arup Partners
STRUCTURAL SYSTEM: Mega core system/reinforced concrete

(credit for photo: CTBUH)

Aspire Tower: architectural and structural information

The 36-storey, 300 m high Aspire Tower in Doha (Qatar) was designed by Hadi Simaan. It is a reinforced concrete building with a mega core system.

The Aspire Tower, whose shape represents a hand holding a flaming torch, was the most important building of the 15th Asian Games held in Qatar in 2006 (Figure 4.8). The 238 m high mega core, together with the 62 m high steel truss structure it carries, completes the design of the building.

The mega core as the spine of the building resists all the entire vertical and lateral loads and supports the cantilevered modules of the building together with the curved external facade. A reinforced concrete mega core shear wall has a circular cross-section with an external diameter varying between 18 to 13 m (from bottom to top) and thickness varying between 2 to 1 m (from bottom to top) throughout the height of the building (Figure 4.9).

FIGURE 4.8 Aspire Tower
(credit for photos: MIDMAC Contracting Co. WLL)

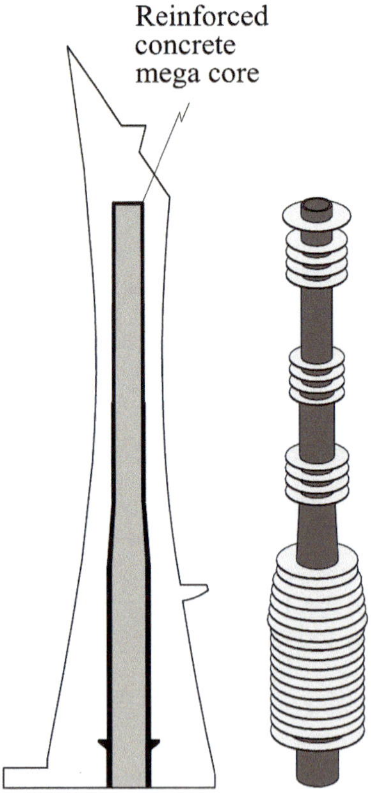

FIGURE 4.9 Aspire Tower structural system

Composite (metal deck with concrete topping) floor slabs in the modules are cantilevered up to 11.3 m from the core and supported by discontinuous steel perimeter columns down through the height of the modules. The bottom slab of each module is a strengthened cantilever floor slab which supports the perimeter columns of the upper storeys in the module. Strengthened cantilever slabs of modules protrude from

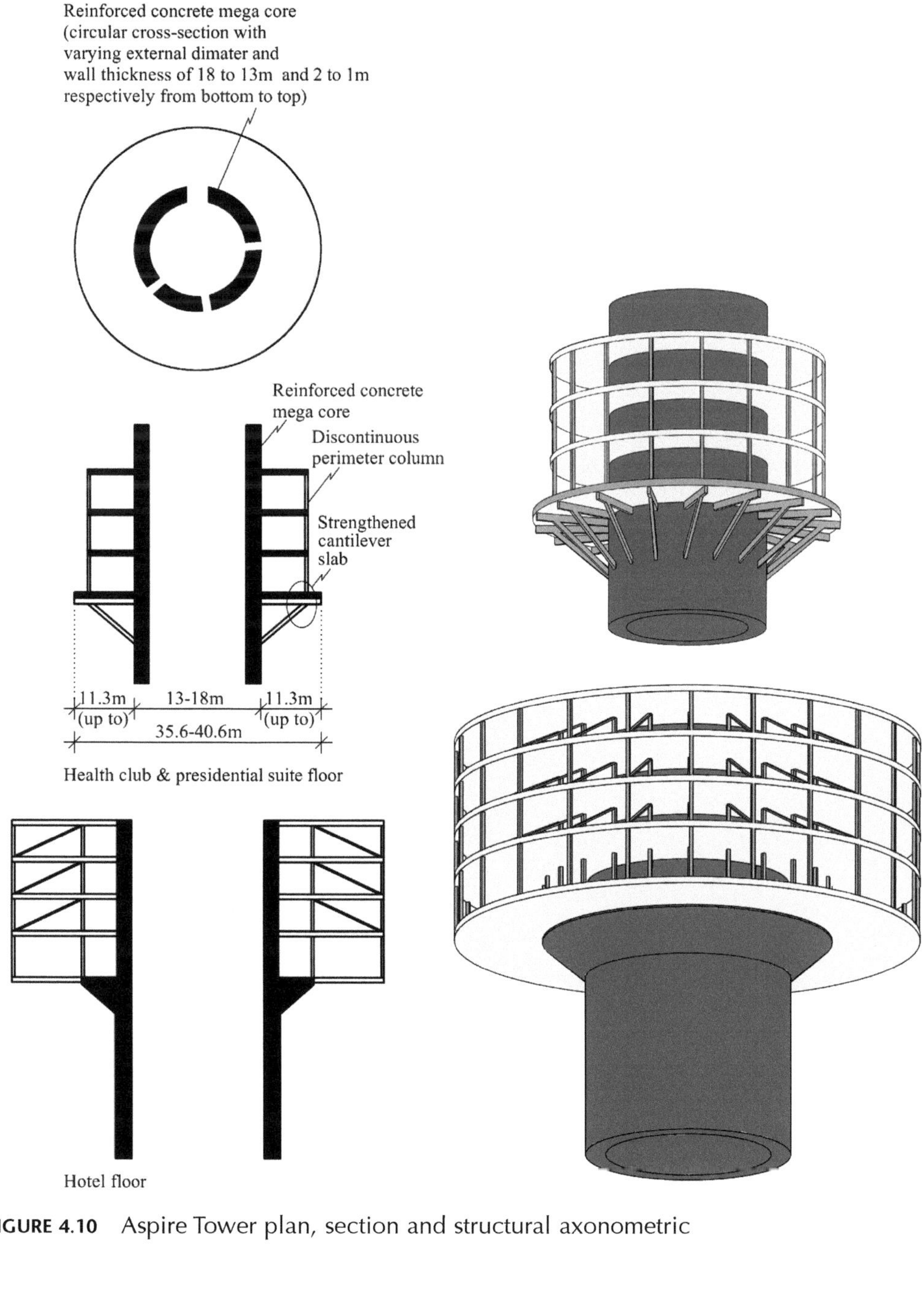

FIGURE 4.10 Aspire Tower plan, section and structural axonometric

the core and are supported at the cantilever root by steel radial beams with brackets (Figure 4.10).

Some part of the surface of the facade on the building is in the form of permeable mesh and some part being in the form of solid cladding. By means of the wind permeable part of the facade, the across-wind effect (wind induced turbulence force or vortex shedding force) on the building is reduced and as a result, the response of the building in the along-wind direction, rather than its response in the across-wind direction (which is generally more complicated and critical than the along-wind response with regard to the building acceleration and occupant comfort) becomes critical and governs the design.

Bibliography

Aspire Tower, http://www.ctbuh.org, accessed May 2012.
Aspire Tower, http://www.emporis.com, accessed May 2012.
Chikaher, G. and Hirst, J., Aspire Tower, Doha, Qatar, *The Arup Journal* 2/2007, pp. 3–13, 2007.
Lubell, S., Sports City Tower, Doha, Qatar, *Architectural Record*, August 2007.

HSB Turning Torso

OFFICIAL NAME: HSB Turning Torso
LOCATION: Malmö, Sweden
BUILDING FUNCTION: Residential
ARCHITECTURAL HEIGHT: 190 m
NUMBER OF STOREYS: 57
STATUS: Completed
COMPLETION: 2005
ARCHITECT: Santiago Calatrava
STRUCTURAL ENGINEER: SWECO
STRUCTURAL SYSTEM: Mega core system/reinforced concrete

(photo on left courtesy of PERI Gmbh and photo on right courtesy of Santiago Calatrava/ Samark Architecture & Design)

HSB Turning Torso: architectural and structural information

The 57-storey, 190 m high HSB Turning Torso in Malmö (Sweden) was designed by Santiago Calatrava. It is a reinforced concrete building with a mega core system. The HSB Turning Torso was awarded a prize by the International Concrete Federation as "the world's most technically interesting and spectacular reinforced concrete building constructed in the last 4 years".

The HSB Turning Torso, is an important project in the redevelopment plan for the residential zone in the industrial district known as "the Western Harbour". The

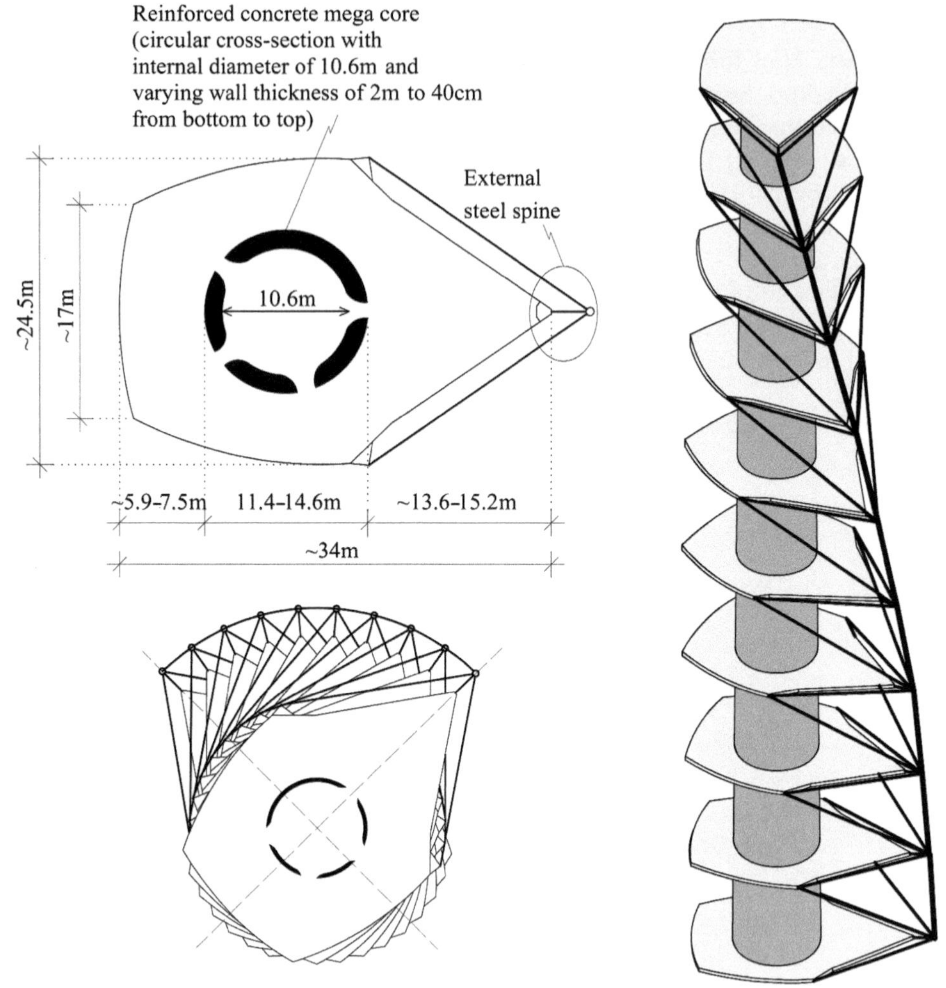

FIGURE 4.11 HSB Turning Torso plans and structural axonometric

building, which has become an icon for Malmö, was designed as a response to the need for housing.

The well-known Spanish architect/engineer Santiago Calatrava, the designer of the building, was inspired in designing the HSB Turning Torso by his own sketch entitled "Twisting Torso". The sketch, which depicts a turning human body (torso), guided the form of the building and consists of 9 modules positioned on top of one another, with a facade twisting through 90° from bottom to top. The HSB Turning Torso is both the first residential building and the first tall building to be designed by Santiago Calatrava.

The HSB Turning Torso consists of a central mega core and nine 5-storey modules having pentagonal floor shape. Mechanical equipment floors are located in the 2 m deep gap spaces between the modules.

The central mega core supports the entire vertical and lateral loads. It is a reinforced concrete core shear wall having circular cross-section with an internal diameter of 10.6m and wall thickness varying from 2m to 40cm from bottom to top so that its external diameter varies between 14.6 to 11.4m (from bottom to top) (Figure 4.11). In addition to the mega core, a reinforced concrete perimeter column and an exoskeleton (an exterior truss), both with the same rotation as the tower, are located at the tip of the triangular part of the floor slabs. The exoskeleton is attached to the modules by horizontal and diagonal steel members (Figure 4.11). Both the perimeter column and the exoskeleton not only help to support the cantilevered floor slabs, but contribute positively to the central core by reducing the lateral drift of the building created by wind loads.

Three edges of the pentagonal shaped floor slabs are slightly curved and the other two edges forming the apex of the pentagon are straight. Reinforced concrete floor slabs in the modules are cantilevered from the core and are supported by discontinuous steel perimeter columns down through the height of the modules. The bottom slab of each module is a strengthened cantilever floor slab which supports the perimeter columns of the upper storeys in the module. While floor slabs are 27cm thick, strengthened cantilever slabs of modules that protrude from the core are 90cm thick at the cantilever root reducing to 40cm at the perimeter.

Bibliography

From Twisting to Turning Torso, *Kinnarps Magazine*, No. 5, 2006.

Reina, P., *Architect-Engineer Santiago Calatrava Deftly 'Twists' Tower*, ENR, December 13, 2004.

Turning Torso, *Yapı Dergisi*, No. 290, pp. 54–59, January 2009. (Translated from English by Burçin Yılmaz)

Turning Torso, *Architectural Record*, Building Types Study – Tall Buildings, August 2007.

Turning Torso, http://www.ctbuh.org, accessed May 2012.

Turning Torso, http://www.emporis.com, accessed May 2012.

Burj Khalifa

OFFICIAL NAME: Burj Khalifa (formerly Burj Dubai)
LOCATION: Dubai, U.A.E
BUILDING FUNCTION: Mixed-use (hotel, residential, office)
ARCHITECTURAL HEIGHT: 828 m
NUMBER OF STOREYS: 163
STATUS: Completed
COMPLETION: 2010
ARCHITECT: Skidmore, Owings & Merrill – SOM; Hyder Consulting
STRUCTURAL ENGINEER: William F. Baker (SOM)
STRUCTURAL SYSTEM: Outriggered frame system/reinforced concrete

(photos courtesy of Adrian Peret, adrian.peret@gmail.com)

Burj Khalifa: architectural and structural information

The 163-storey, 828 m high Burj Khalifa in Dubai (U.A.E) was designed by SOM and Hyder Consulting. It is a reinforced concrete building with an outriggered frame system. The system is also classified as a buttressed core system. The Burj Khalifa gained the title of "the world's tallest building" in 2010. Moreover, it was the winner of "Global Icon Award 2010" by CTBUH; "Best Tall Building 2010, Middle East and Africa" by CTBUH; and "Distinguished Building Award" in 2011 by AIA (American Institute of Architects).

The design of the Burj Khalifa, which derives its main inspiration from the form of a local desert flower, has many traditional Islamic architectural motifs. The cross-section has three axes in the shape of a Y, whose wings end by evoking the shape of a dome. The Y-shaped plan form maximises the exterior view and secures a supported core which is termed "buttressed core". The building has the appearance of bundled modules that end at different levels so that 21 setbacks are created in a spiral manner below the spire part (structural steel braced frame) of the building. Although it is reminiscent of the bundled-tube form it does not have a tubular structural system. The reduction of the plan area by using setbacks throughout the height of the building, while further highlighting the height of the tower, which appears to be rising towards the sky with acceleration, also reduces wind forces at the upper levels by a reduction in the surface area affected by the wind.

According to statements by representatives (William F. Baker, Peter A. Irwin and Ahmad Abdelrazag) of the firms involved in the architectural and structural design of the Burj Khalifa (Skidmore, Owings & Merrill), the wind engineering consultancy (Rowan Williams Davies and Irwin) and the construction (Samsung Engineering and Construction), attention was paid to the effect of wind and aerodynamic form at an early stage in the architectural design and the architectural and structural designs were undertaken in parallel. Thus the Burj Khalifa took shape with the collaborative efforts of architects and engineers in order to reduce the effect of wind as much as possible. The number and location of the modules formed by setbacks throughout the height of the building and the shape of the wings, were determined by wind tunnel tests. The setbacks on the building reduce the effect of wind induced lateral loads by breaking the organisation of the wind flow, namely confusing the wind because reduced plan area especially in a spiral manner causes the wind to encounter reduced and different surface areas throughout the height of the building.

The structural system of the Burj Khalifa is composed of a hexagonal central core (buttressed by wing shear walls) and outriggers. Each wing has hammerhead ended corridor shear walls (extending from the central core), perimeter shear walls at the sides and circular nose columns at the tip. Outriggers connect the core with the perimeter shear walls and nose columns through buttresses (wing shear walls). Multi-storey outrigger shear walls at the mechanical floors at 5 levels increase the strength of the buttressed core against lateral loads and thus the structural system is an outriggered frame system (outriggered frame system with buttressed core) (Figure 4.12). The hexagonal central core consists of reinforced concrete shear walls with thicknesses varying between 130 cm at the bottom to 50 cm at the top (below the spire) through the height of the building.

Wind force was dominant in the lateral loading and it was accepted that the maximum lateral drift at the top of the building would be 1.2 m. The setbacks and wings on the building were developed using wind tunnel tests on a 1:500 scale model and at every stage the form of the building was re-shaped after repeating these tests, which resulted in a reduction of the wind load to an absolute minimum.

High strength concrete was used in the Burj Khalifa, varying in strength between 80 MPa and 60 MPa throughout the height of the building from bottom to top.

There is a 232 m high steel structure in the upper part of the building, consisting of brace elements and with self-resistance against vertical and lateral loads, which is supported by the central core

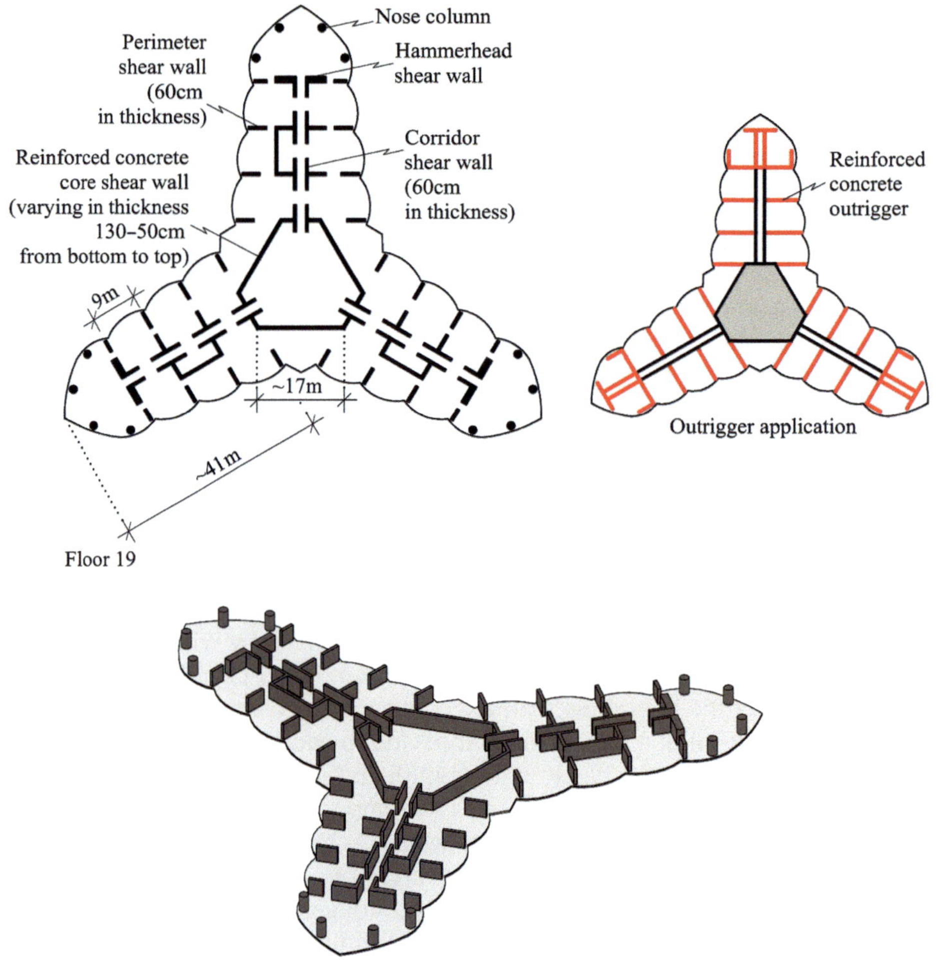

FIGURE 4.12 Burj Khalifa plan and structural axonometric

The slab system on each storey consists of two-way reinforced concrete flat plates that vary between 20 and 30 cm in depth as they pass through spaces of approximately 9 m between the nose columns, perimeter shear walls and the hexagonal central core, while on the upper storeys they vary between 22.5 and 25 cm in depth.

Bibliography

Abdelrazaq, A., Validating the Dynamics of the Burj Khalifa, *CTBUH Journal*, CTBUH Technical Paper, Issue II, 2011.

Abdelrazaq, A., Kim, K.J. and Kim, J.H., *Brief on the Construction Planning of The Burj Dubai Project,* UAE, CTBUH 8th World Congress 2008, Dubai, 2008.

Baker, W., *The World's Tallest Building, Burj Dubai, UAE,* Proceedings of the CTBUH 2004 Seoul Conference, pp. 1168–1169, Seoul, 2004.

Baker, W.F., Korista, D.S. and Novak, L.C., *Engineering the World's Tallest – Burj Dubai,* Proceedings of the CTBUH 8th World Congress 2008, Dubai, 2008.

Baker, W.F., Korista, D.S. and Novak, L.C., Burj Dubai: Engineering the World's Tallest Building, *The Structural Design of Tall and Special Buildings*, Vol. 16, pp. 361–375, 2007.
Burj Dubai, http://www.ctbuh.org, accessed May 2012.
Burj Dubai, http://www.emporis.com, accessed May 2012.
Irwin, P.A., *The Wind Engineering of the Burj Dubai Tower*, CTBUH 7th World Congress 2005, New York, 2005.
Irwin, P.A. and Baker, W.F., The Burj Dubai Tower: Wind Engineering, *Structure Magazine*, pp. 28–31, June 2006.
Irwin, P.A., Baker, W.F., Korista, S., Weismantle, P.A. and Novak, L.C., The Burj Dubai Tower: Wind Tunnel Testing of Cladding and Pedestrian Level, *Structure Magazine*, pp. 48–51, November 2006.
Smith, A., *Burj Dubai: Designing the World's Tallest*, UAE, CTBUH 8th World Congress 2008, Dubai, 2008.
Weismantle, P.A., Smith, G.L. and Sheriff, M., Burj Dubai: An Architectural Technical Design Case Study, *The Structural Design of Tall and Special Buildings*, Vol. 16, pp. 335–360, 2007.

Taipei 101

OFFICIAL NAME: Taipei 101
LOCATION: Taipei, Taiwan
BUILDING FUNCTION: Office
ARCHITECTURAL HEIGHT: 508 m
NUMBER OF STOREYS: 101
STATUS: Completed
COMPLETION: 2004
ARCHITECT: C.Y. Lee & Partners
STRUCTURAL ENGINEER: Thornton Tomasetti; Evergreen Engineering
STRUCTURAL SYSTEM: Outriggered frame system/composite

Taipei 101: architectural and structural information

The 101-storey, 508 m high Taipei 101 in Taipei (Taiwan) was designed by C.Y. Lee & Partners. It is a composite building with an outriggered frame system. Its aspect ratio is approximately 9. The Taipei 101 gained the title of "the world's tallest building" in 2004 and it is the first building to have exceeded the half kilometre limit.

The Taipei 101, with a design inspired by the form of bamboo, has a ground floor with a 63.5×63.5 m square cross-section, then a 25-storey pyramid form on top of which are 8 modules consisting of 8-storey truncated inverted pyramids (slightly outward sloping towards the top). The top of the building is in the shape of a 12-storey truncated inverted pyramid. According to Chinese belief, the number 8 is identified with wealth and abundance. The floor-to-floor height is 4.2 m.

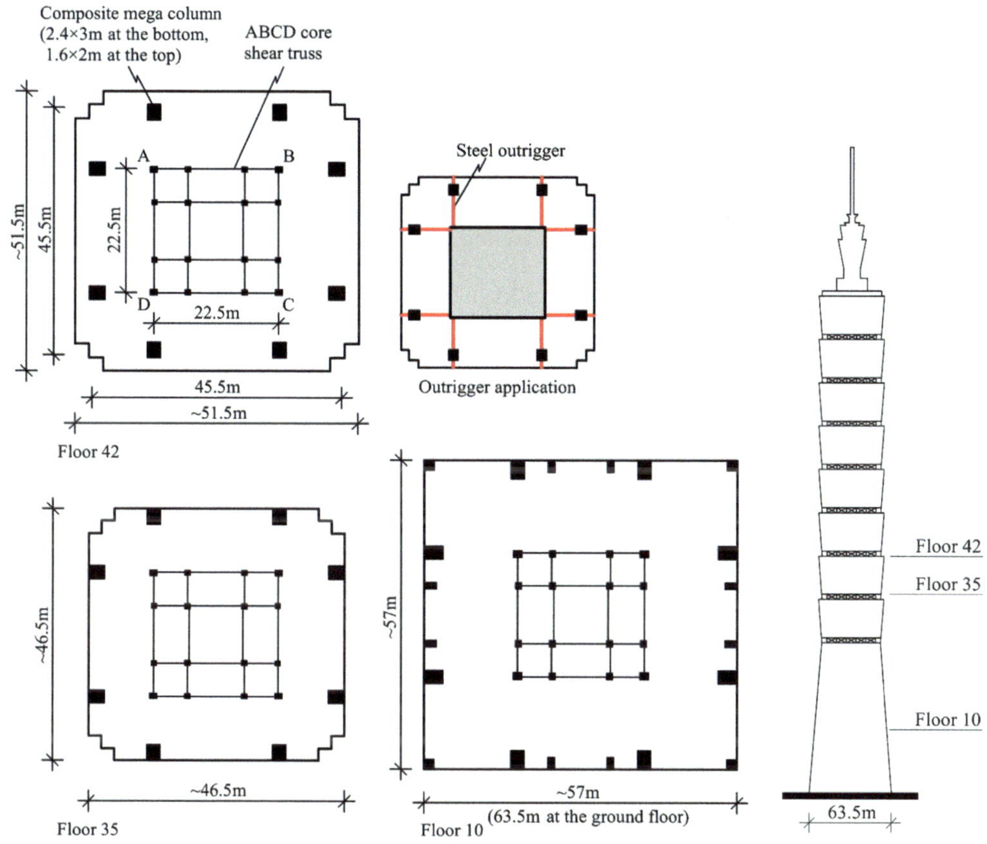

FIGURE 4.13 Taipei 101 schematic section and plans

The Taipei 101 is an example of a building with setbacks and an aerodynamic building top (Section 6.1). In addition, the application of the saw-tooth corner, a development of the recessed corner, to the building significantly reduces wind loads on the building, compared with a sharp corner and reduces wind induced base moment by 25 per cent (Section 6.1). In the design of the Taipei 101 structural system, resistance against the wind, which is estimated to be capable of reaching a speed of 43.3 m/s (156km/h), was an important design input.

The 8 perimeter columns and 16 core columns, all composite, consist of box-section steel filled with high-strength concrete (70 MPa). At the ground floor the perimeter columns have dimensions of 2.4×3 m. The perimeter and core columns are connected by outriggers, 1 or 2 storeys deep, at 10 levels along the height of the building (Figure 4.13).

The Taipei 101 has a lateral drift limit ratio of 1/200 of its height and a 730-ton tuned mass damper (TMD) was used near the top of the building (between the 87th and 92nd floors) (Figure 6.19).

Bibliography

Hadenius, P., The World's Tallest Building for Now, *Technology Review*, Massachusetts Institute of Technology/MIT, August 2004.

Joseph, M.L., Poon, D.C.K. and Shieh, S., Taipei 101: The World's Tallest Building, *Structure Magazine*, pp. 40–45, June 2006.

Poon, D.C.K., Shieh, S., Joseph, M. L. and Chang, C., *Structural Design of Taipei 101: The World's Tallest Building*, CTBUH Seoul Conference, pp. 271–278, Seoul, 2004.

Poon, D.C.K., Shieh, S., Joseph, M.L. and Chang, C., Reaching for the Sky, *Civil Engineering*, pp. 56–62, January 2004.

Taipei 101, http://www.ctbuh.org, accessed May 2012.

Taipei 101, http://www.emporis.com, accessed May 2012.

Taranath, B., *Wind and Earthquake Resistant Buildings: Structural Analysis and Design*, A Series of Reference Books and Textbooks (Editor: Michael D. Meyer), Department of Civil and Environmental Engineering, Georgia Institute of Technology, Atlanta, Georgia, 2005.

Shanghai World Financial Center (SWFC)

OFFICIAL NAME: Shanghai World Financial Center
LOCATION: Shanghai, China
BUILDING FUNCTION: Mixed-use (office, hotel)
ARCHITECTURAL HEIGHT: 492 m
NUMBER OF STOREYS: 101
STATUS: Completed
COMPLETION: 2008
ARCHITECT: Kohn Pedersen Fox Associates; Irie Miyake Architects and Engineers
STRUCTURAL ENGINEER: Leslie E. Robertson Associates
STRUCTURAL SYSTEM: Outriggered frame system/composite

(photos courtesy of Niels Jakob Darger)

Shanghai World Financial Center: architectural and structural information

The 101-storey, 492 m high Shanghai World Financial Center (SWFC) in Shanghai (China) was designed by Kohn Pedersen Fox Associates and Irie Miyake Architects and Engineers (Figure 4.14). It is a composite building with an outriggered frame system. Its aspect ratio is approximately 8.5. The SWFC was named as the winner of "Best Tall Buildings 2008, Asia and Australasia" by CTBUH.

The SWFC has a square cross-section at ground level. The cross-section changes from a square to a hexagon throughout the height of the building, ending at the top of the building as a rectangle. At the point where it changes from a hexagon to a rectangle, at two opposite sides of the building, sloping towards the central core as it approaches the top of the building there is thus a reduction in the cross-sectional area (Figures 4.14–4.15).

The circle and square, symbolising heaven and earth respectively according to Chinese belief, were prominent in the architectural design. There was initially a circular opening at the top of the building, which itself is prismatic with a square base. Later, the circular opening at the top of the building was replaced by a trapezoidal opening.

The SWFC's form, which narrows as it rises towards the top, has various advantages. A continuous relationship between form and function is achieved by locating banking and finance units, which need wide spaces and openings, on the lower floors and a hotel, which needs smaller spaces and openings, on the upper floors. By reducing the plan area, the surface area affected by the wind on the upper levels of the building is reduced, as is the wind intensity and thus the excess wind pressure (Chapter 6). In calculating the aspect ratio (the ratio of the structural height of a building to the narrowest structural width on the floor plan), because the widest dimensions of the floor plan of the building were taken as a basis, the wide ground floor layout improves this ratio and thus lengthens this slender and flexible building (Figures 4.14–4.15).

Concerns over the aerodynamic building top played an important role in the architectural design of the SWFC (Chapter 6).

The structural system of the building was designed as a shear walled frame system and the foundation construction was completed. After the completion of the pile foundation construction, the decision was taken to increase the height from 460 to 492 m and the sides of the square cross-section of the ground floor from 55.8 to 58 m. To prevent the load-carrying capacity of the building foundations from being insufficient due to the negative effect of the increase in height, there was a need to reduce the weight of the building by 10 per cent. In order to lighten the building, the thickness of the shear walls was reduced, because the reinforced concrete core shear wall had the largest share of the total weight. This solution made it necessary to increase the structure's lateral stiffness because it reduced the resistance of the core against wind and earthquake induced lateral loads, thus reducing the lateral stiffness of the building. Thus, mega braces and belts to support the columns and outriggers to support the core were added to the system and what was initially intended to be a shear walled frame system was turned into an outriggered frame system.

In the structural system, there are a reinforced concrete core shear wall, composite mega columns, composite outriggers, steel belts and composite mega braces (Figures 4.15–4.16).

Composite mega columns are located on the corners. They are composite elements made of structural steel sections encased in reinforced concrete. The columns have pentagonal cross-sections, with the longest two sides being 5.45 m at the ground floor (Figure 4.17).

Composite mega braces are 12-storey-high and are composed of box-section steel filled with reinforced concrete.

Outriggers are composed of 3-storey-deep composite trusses connecting the mega columns to the reinforced concrete core shear wall.

Belts, consisting of 1-storey-deep steel trusses, are located at every 12 floors.

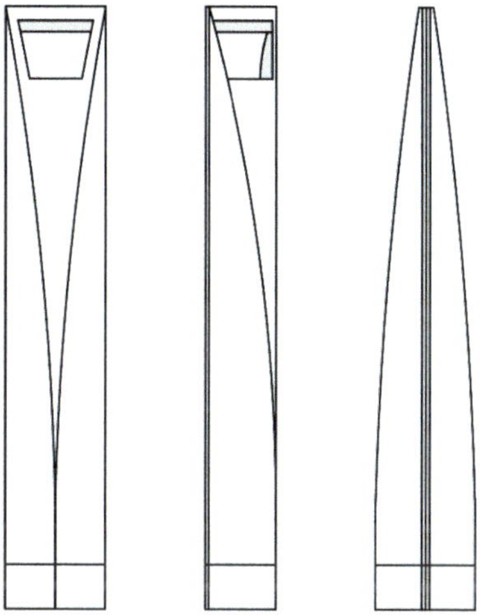

FIGURE 4.14 SWFC schematic elevations

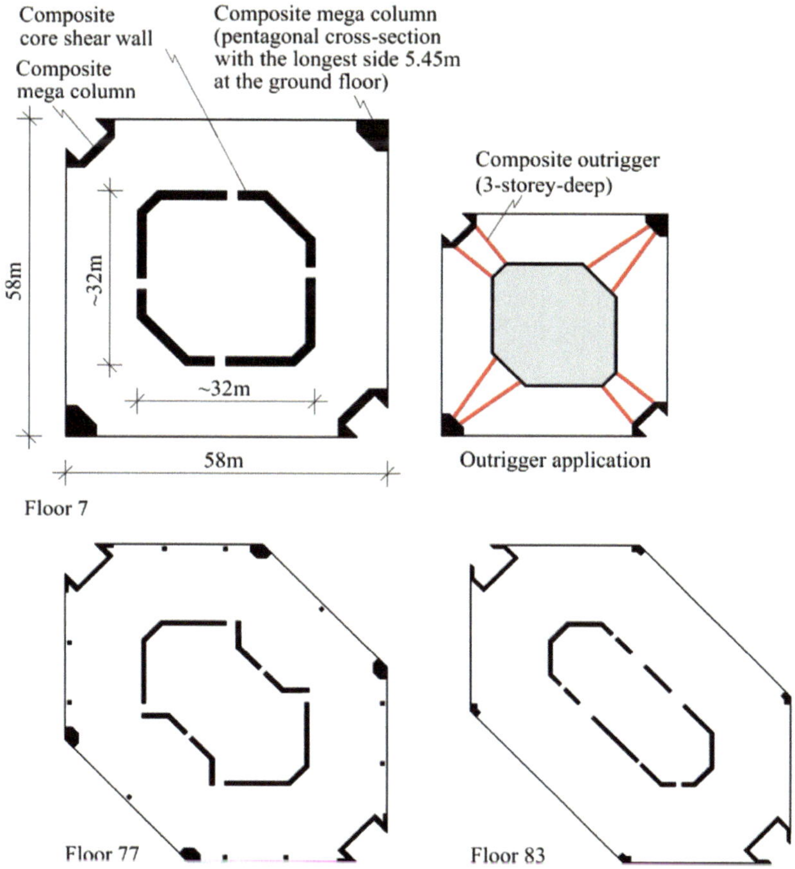

FIGURE 4.15 SWFC plans

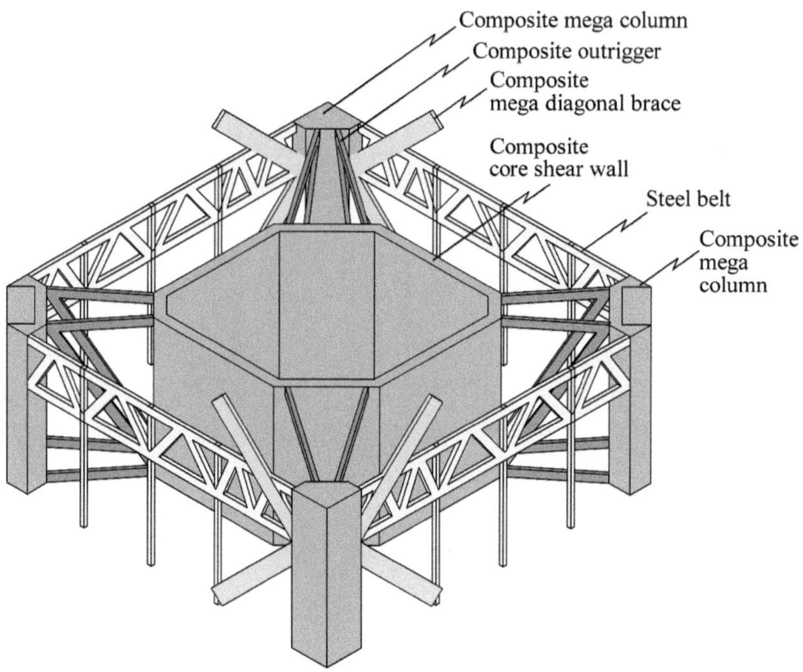

FIGURE 4.16 SWFC structural axonometric

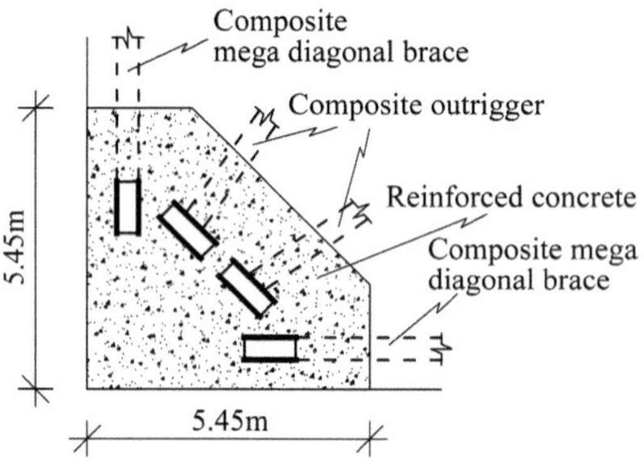

FIGURE 4.17 SWFC composite mega column section

Bibliography

Chen, A., Shanghai World Financial Center, *Architectural Record* 05.09, pp. 184–190, 2009.

Katz, P., Robertson, L.E. and See, S., *Shanghai World Financial Center: Without Compromise*, CTBUH 8th World Congress 2008, Dubai, 2008.

Katz, P. and Robertson L.E., Case Study: Shanghai World Financial Center, *CTBUH Journal*, No. 2, 2008.

Lu, X., Zou, Y., Lu, W. and Zhao, B., Shaking Table Model Test on Shanghai World Financial Center Tower, *Earthquake Engineering and Structural Dynamics*, Vol. 36, pp. 439–457, 2007.

Robertson, L.E. and See, S., The Shanghai World Financial Center, *Structure Magazine*, pp. 32–35, June 2007.

Shanghai World Financial Center, http://www.ctbuh.org, accessed May 2012.

Shanghai World Financial Center, http://www.emporis.com, accessed May 2012.

The Petronas Twin Towers

OFFICIAL NAME: The Petronas Twin Towers
LOCATION: Kuala Lumpur, Malaysia
BUILDING FUNCTION: Office
ARCHITECTURAL HEIGHT: 452 m
NUMBER OF STOREYS: 88
STATUS: Completed
COMPLETION: 1998
ARCHITECT: César Pelli & Associates
STRUCTURAL ENGINEER: Thornton Tomasetti; Ranhill Bersekutu
STRUCTURAL SYSTEM: Outriggered frame system/reinforced concrete

The Petronas Twin Towers: architectural and structural information

The 88-storey, 452 m high Petronas Twin Towers in Kuala Lumpur (Malaysia) were designed by César Pelli & Associates. They are reinforced concrete buildings with an outriggered frame system. Their aspect ratio is approximately 8.6. The Petronas Twin Towers gained the title of "the world's tallest buildings" in 1998.

The Petronas Twin Towers were the first "world's tallest building" to be built outside the United States of America and also the first skyscrapers in the race for height to be made of reinforced concrete, rather than steel. César Pelli said that he had tried to express the essence of Malaysia, its richness in culture and its extraordinary vision for the future and that the building was rooted in tradition and was about Malaysia's aspiration and ambition.

The 8-pointed star formed from two overlapping squares and 8 semi-circles provided the inspiration for the design of the Petronas Twin Towers (Figure 4.18). The surface area of the buildings was increased using this cross-section, which expresses architecturally important Islamic principles such as "unity, harmony, stability and rationality", giving an unusual and impressive complexity to the essentially plain appearance of the facade and maximising the view from inside.

The Petronas Twin Towers are an example of buildings where the plan area has been reduced by setbacks on the facade and that have an aerodynamic top (Figure 4.19) (Section 6.1).

The Petronas Twin Towers' two identical cylindrical towers, with a diameter of 46.3 m, have a floor-to-floor height of 4 m.

Three of the five facade setbacks (on the 60th, 73rd and 82nd floors) consist of 3-storey-high sloping columns. A 58.4 m steel skybridge supported by arches between the 41st and 42nd floors "symbolizes the threshold between the tangible and the spiritual worlds", according to César Pelli (Figure 4.20).

The diameter of the 16 perimeter columns varies from 240 cm at the bottom to 120 cm at the top. The columns are spaced at approximately 9 m centres. Reinforced concrete cores of the towers, beginning with square cross-sections of 22.9×22.9 m at the base, decrease in size towards the top of the buildings in steps, ending with rectangular cross-sections of 18.9×22 m.

Reinforced concrete cores have inner walls with a thickness of 35 cm and outer walls varying in thickness from 75 to 35 cm from bottom to top. Ring beams are 79 cm

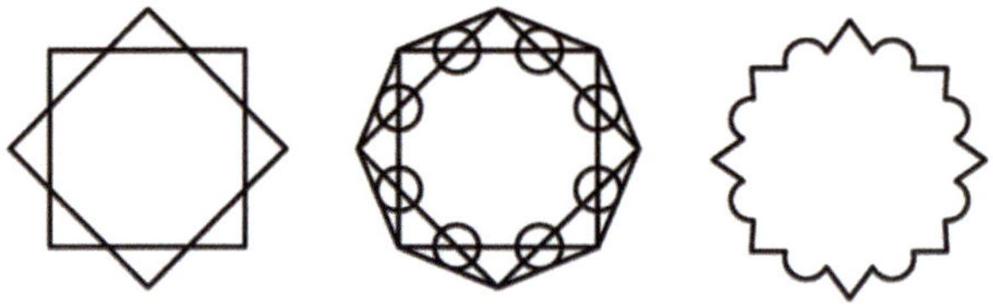

FIGURE 4.18 The Petronas Twin Towers plan evolution

FIGURE 4.19 The Petronas Twin Towers, aerodynamic top
(photo courtesy of Antony Wood/CTBUH)

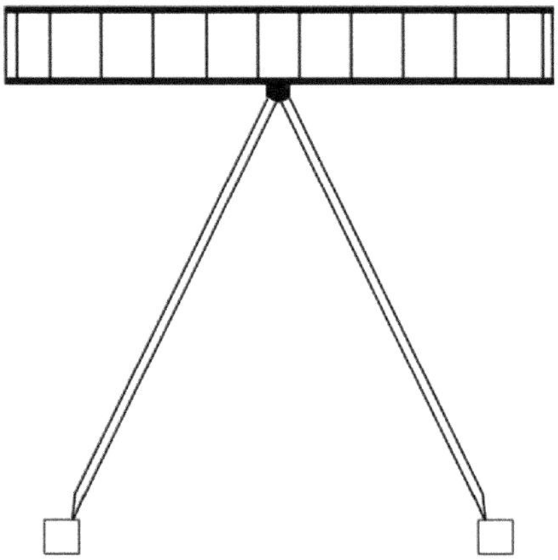

FIGURE 4.20 The Petronas Twin Towers skybridge

deep at midspan and 117 cm deep at column faces. For additional stiffness, reinforced concrete core shear walls and perimeter columns are connected to each other by 2-storey-deep outriggers at the 38th floor (floors 38–40 – at almost mid-height of the structure), which is a mechanical equipment floor (Figure 4.21).

High-strength concrete was used in the building, 80 MPa in the lower floors, 60 MPa in the middle floors and 40 MPa in the upper floors.

The Petronas Twin Towers' composite floor slabs, supported by wide-flange beams, consist of 5.1 cm deep trapezoidal metal deck covered with a layer of reinforced concrete with a thickness of 11.4 cm in the office floors and 20.3 cm at the mechanical equipment floors (Figure 4.21).

Bibliography

Ali, M. and Armstrong, P., *Architecture of Tall Buildings*, Council on Tall Buildings and Urban Habitat Committee (CTBUH), New York: McGraw-Hill Book Company, 1995.

Ho, P.H.K, *Economics Planning of Super Tall Buildings in Asia Pacific Cities*, Strategic Integration of Surveying Services FIG Working Week 2007, Hong Kong, China, 13–17 May 2007.

Kareem, A., Kijewski, T. and Tamura, Y., Mitigation of Motion of Tall Buildings with Specific Examples of Recent Applications, *Wind and Structures*, Vol. 2, No. 3, pp. 201–251, 1999.

Kim, J.O. and Lee, S.H., *Application of High Performance Concrete in Petronas Twin Tower, KLCC*, CTBUH Seoul Conference, pp. 457–464, Seoul, 2004.

Naidu, K.G., *The Petronas Tower: The World's Tallest Building*, CPAC Seminar 95, Modern Technology in Concrete Construction, Bangkok, Thailand, 1995.

Pollalis, S., *The Petronas Towers*, Harvard Design School, Center for Design Informatics, 2002.

Taranath, B., *Steel, Concrete & Composite Design of Tall Buildings*, New York: McGraw-Hill Book Company, 1998.

The Petronas Twin Towers, http://www.ctbuh.org, accessed May 2012.

The Petronas Twin Towers, http://www.emporis.com, accessed May 2012.

Thornton, C., Hungspruke, U. and Joseph, L., Design of the World's Tallest Buildings – Petronas Twin Towers at Kuala Lumpur City Center, *The Structural Design of Tall Buildings*, Vol. 6, pp. 9–27, 1997.

Zaknic, I., Smith, M. and Rice, D., *100 of The World's Tallest Buildings*, Council on Tall Buildings and Urban Habitat Committee, New York: McGraw-Hill Book Company, 1998.

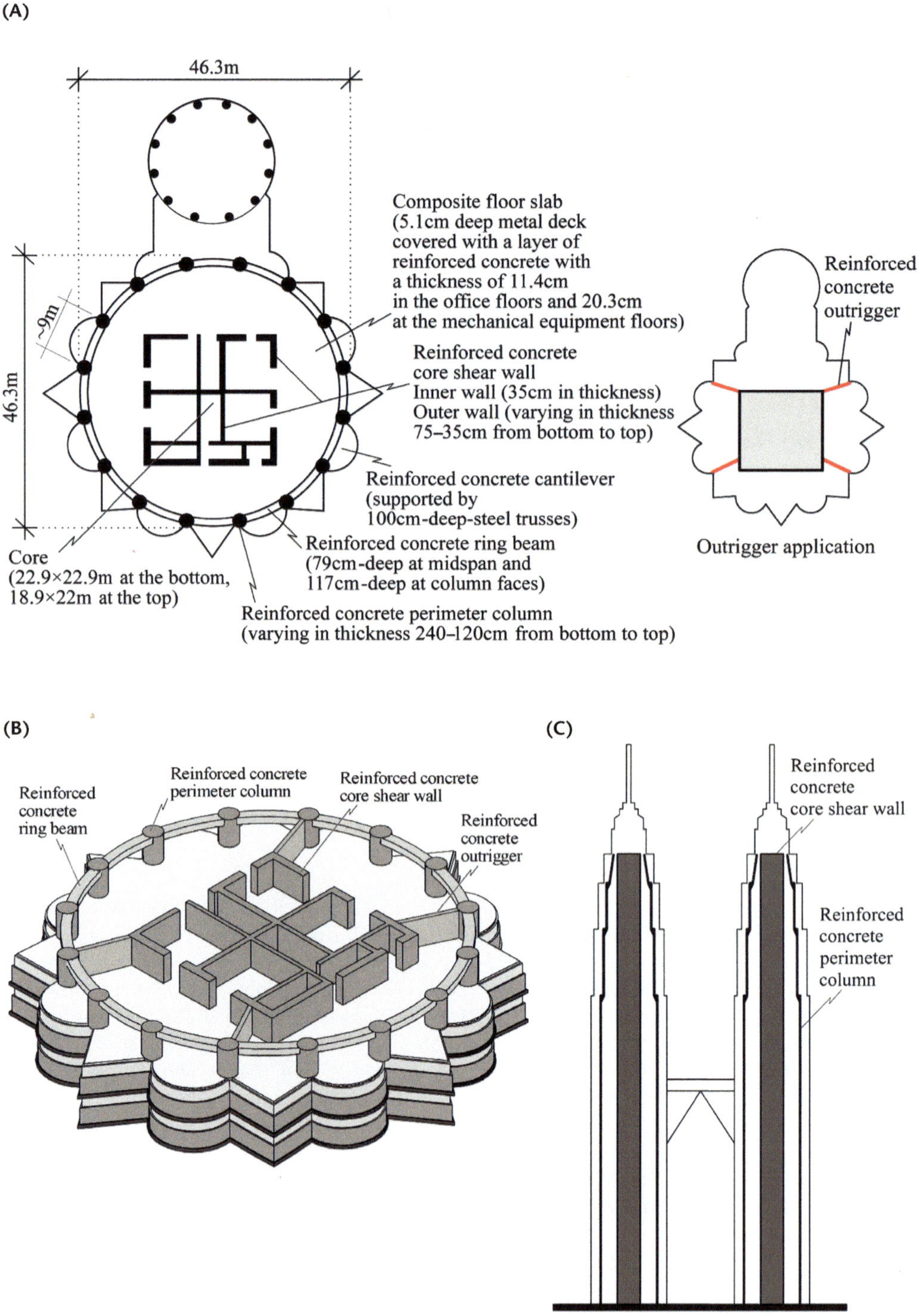

FIGURE 4.21 The Petronas Twin Towers: (a) plan, (b) structural axonometric, (c) schematic section

New York Times Tower

OFFICIAL NAME: New York Times Tower
LOCATION: New York, USA
BUILDING FUNCTION: Office
ARCHITECTURAL HEIGHT: 319 m
NUMBER OF STOREYS: 52
STATUS: Completed
COMPLETION: 2007
ARCHITECT: FXFOWLE; Renzo Piano Building Workshop
STRUCTURAL ENGINEER: Thornton Tomasetti
STRUCTURAL SYSTEM: Outriggered frame system/steel

(courtesy of Antony Wood/CTBUH)

New York Times Tower: architectural and structural information

The 52-storey, 319 m high New York Times Tower in New York (USA) was designed by FXFOWLE and Renzo Piano Building Workshop. It is a steel building with an outriggered frame system. Its aspect ratio is approximately 6.8. The New York Times Tower gained the title of "the 3rd tallest building in New York" in 2007. It was also chosen as the winner of "Best Tall Building 2008, Americas" by CTBUH; "AIA (American Institute of Architects) New York Chapter Building Type Honor Award 2008, Sustainable Design"; and "AIA Honor Award for Architecture 2009".

The architect Renzo Piano's main idea in the architectural design was a light and transparent building to express the transparency and openness in the journalistic ideal. Renzo Piano says that, "The story of this building is one of lightness and transparency". The resulting design is a building with an inner glazed facade and an outer transparent screen layer composed of closely spaced horizontal ceramic rods mounted outboard of the facade. Ceramic rods are not only features of the architectural design, but are also sun shading without disturbing the transparency of the building. In this way, the transparent double-skin facade system with the shade screen provides the connection between the journalists working in the building and the people outside. Moreover, the transparency of the building is emphasised also by expressing the structure in large corner notches (recessions) by X-braces and columns where the exterior screen layer is omitted.

Above the lobby at the ground floor, floors 2–4 are the newsrooms, 5–27 and 29–50 are for the building occupants. Outriggers and mechanical rooms are located at floors 28 and 51.

The structural core is a centrally located braced core with 27.4×19.8 m in dimensions in the north-south and east-west directions respectively. The least span between the core and the perimeter columns on the long east and west faces is about 12 m. North and south faces have cantilevered floor areas with 19.8×6.1 m in dimensions. Columns on the east and west faces are spaced at 9.14 m centres (Figure 4.22).

Although the structural core area remains constant throughout the building height, above the 27th floor, since the elevators serving the lower half of the building are no longer required, the area left for the service core decreases. This situation changes the bracing configuration of the central core and so increases the space efficiency ratio of the floors in the lower half (floors 5–27) and upper half of the building (floors 29–top) from 0.746 to 0.833 respectively (Figure 4.23).

The Tower has 30 columns extending to the ground level. Exposed columns are made of built-up box sections with dimensions of 76×76 cm. In order to improve the lateral stiffness of the structural system, some of the columns at the lowest floors have solid sections. Inner columns are wide-flange section columns. However, perimeter columns of the cantilevered floor areas do not extend to the ground level.

Two-storey-deep outriggers are located at the mid-height and top mechanical floors of the building. The lateral stiffness of the outriggered frame system is improved by 2-storey-high pretensioned X-braces on the north and south faces of the four notched corners of the building. These braces reduce the maximum lateral drift/sway (drift index) from 1/350 to 1/450 of the building height (90 to 70 cm).

The X-bracing system of the building is very efficient since both the diagonal steel rods are pretensioned so that compression force cannot develop and both the braces work simultaneously.

In general, conventional X-braces are composed of two single diagonal members placed with a certain eccentricity while crossing each other at the middle of the bay (which creates column torsion) (Figure 4.24a), or a crossing joint is placed at the middle of the bay where the braces intersect (to overcome this problem) (Figure 4.24b). However, in the New York Times Building, X-braces are composed of pairs of diagonal members by placing one pair side-by-side and the other pair over-and-under (Figure 4.24c). In this way, the column torsion problem that gains importance because of pretensioning is overcome.

The floor structure of the building is composed of a composite slab system. A 6.4 cm thick topping concrete and 7.6 cm deep trapezoidal metal deck are supported by steel frames.

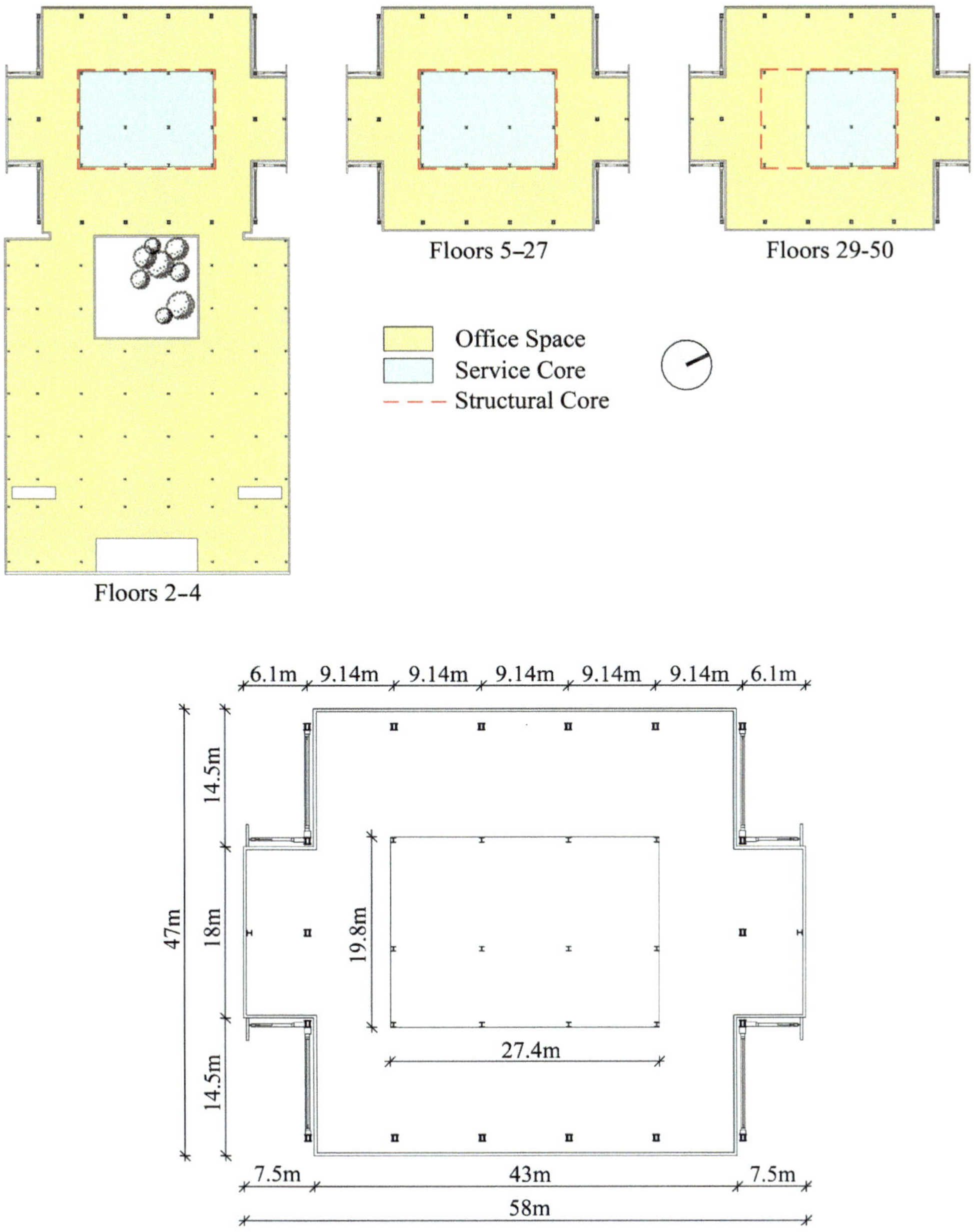

FIGURE 4.22 New York Times Tower plans

(A)

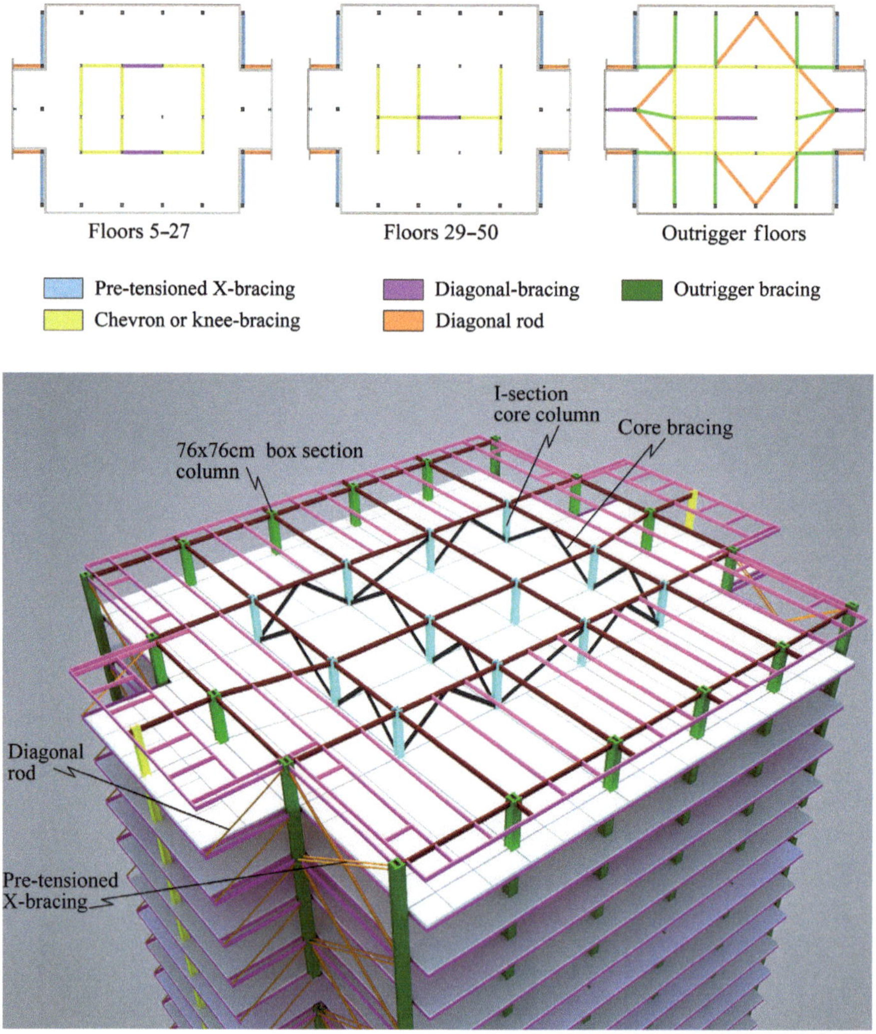

FIGURE 4.23 New York Times Tower: (a) bracing system in plans, (b) structural axonometric (floors 5–27)

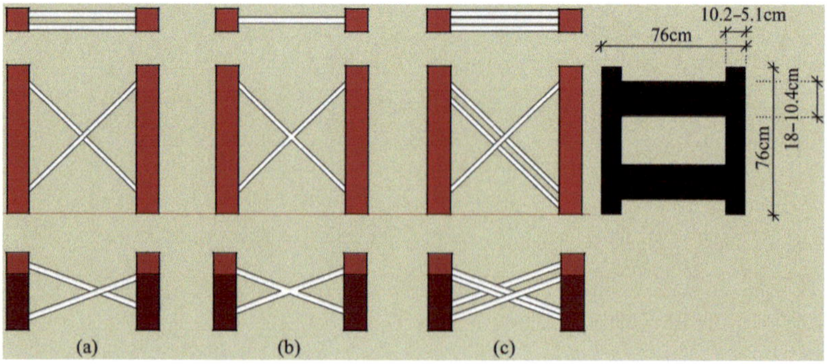

FIGURE 4.24 Detail of X-braced bays on the north and south faces of the notched corners

Bibliography

Agnoletto, M., *Renzo Piano*, Milan: Motta Architettura srl, 2006.
Buchanan, P., *Renzo Piano Building Workshop*, London: Phaidon Press Limited, 2008.
Callow, J.A., Krall, E. and Scarangello, T. Z., *Inside Out*, Modern Steel Construction, January 2009.
Forest City Enterprises, Inc., *Our Newest Value Creation Story: The New York Times Building, 2007 Summary Annual Report and Supplemental Package,* 2007.
Hedrick, M.S., Horst, K., Lehman, C. and Perez, A., *The New York Times Building*, The Pennsylvania State University, Pennsylvania, 2010.
Jodidio, P., *Renzo Piano Building Workshop: 1966 to Today*, Taschen, 2008.
New York Times Tower, http://newyorktimesbuilding.com, accessed May 2012.
New York Times Tower, http://www.ctbuh.org, accessed May 2012.
New York Times Tower, http://www.emporis.com, accessed May 2012.
Scarangello, T.Z., Joseph, L.M. and Krall, K.E., A Statement in Steel: The New York Times Building, *Structural Engineer*, June 2004.
Scarangello, T.Z., Krall, K.E. and Callow, J.A., *A Statement in Steel: The New York Times Building,* CTBUH 8th World Congress, 2008.

Eureka Tower

OFFICIAL NAME: Eureka Tower
LOCATION: Melbourne, Australia
BUILDING FUNCTION: Residential
ARCHITECTURAL HEIGHT: 297 m
NUMBER OF STOREYS: 91
STATUS: Completed
COMPLETION: 2006
ARCHITECT: Fender Katsalidis
STRUCTURAL ENGINEER: Connell Mott MacDonald
STRUCTURAL SYSTEM: Outriggered frame system/reinforced concrete

(photos courtesy of David Randerson)

Eureka Tower: architectural and structural information

The 91-storey, 297 m high Eureka Tower in Melbourne (Australia) was designed by Fender Katsalidis. It is a reinforced concrete building with an outriggered frame system. Its aspect ratio is approximately 7. The Eureka Tower won the title of Best Overend Award of Australian Institute of Architects for Residential Architecture-Multiple Housing in 2007.

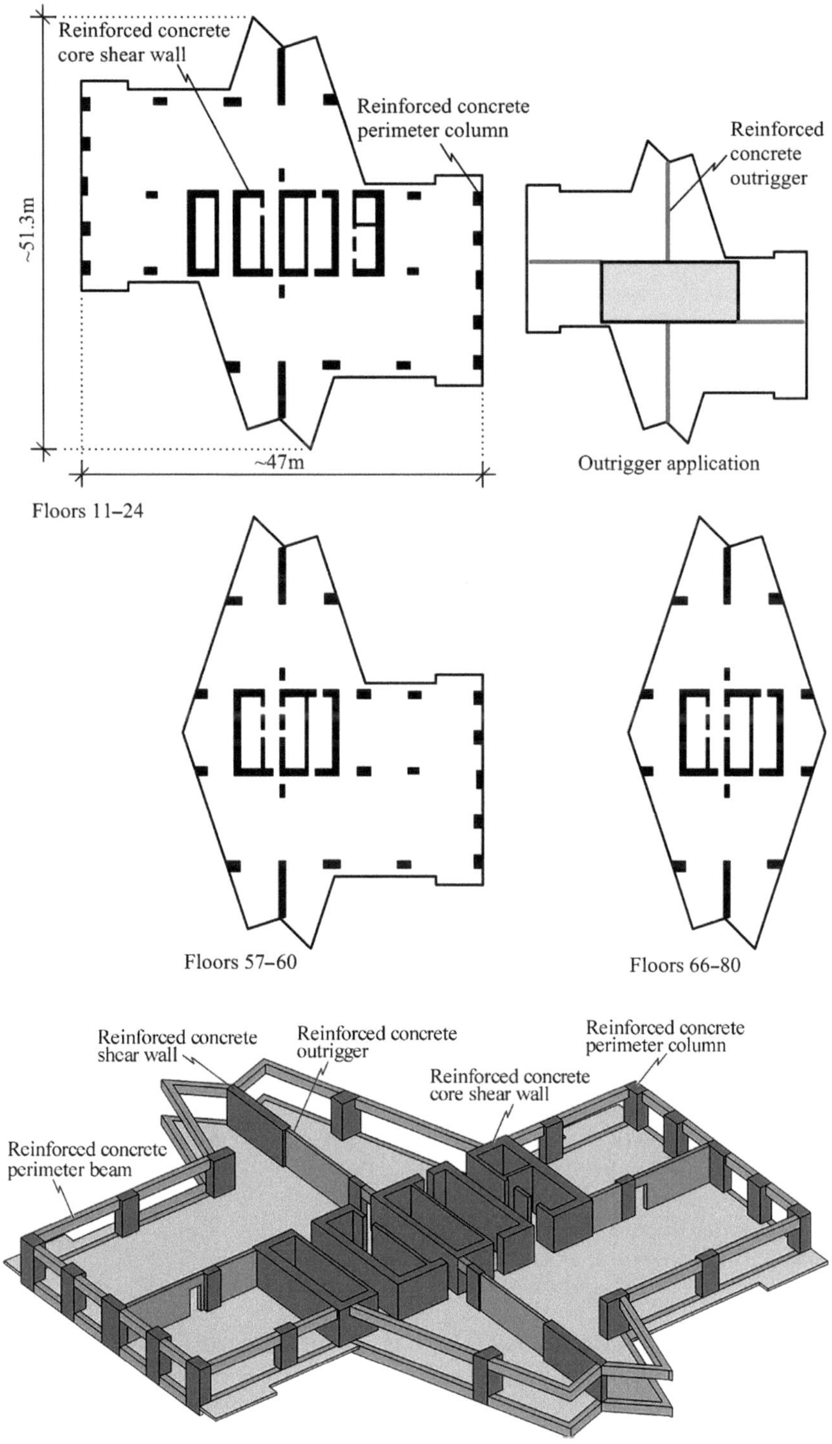

FIGURE 4.25 Eureka Tower plans and structural axonometric

The name Eureka has a special meaning in Australia's past due to the discovery of gold in the nineteenth century. The name of the building is related to the Eureka Stockade, a rebellion of gold miners in 1854, which resulted in many deaths. The flag of the rebels has been incorporated into the design by blue glass cladding and the white lines. The 24 carat gold plated glass windows on the facade of top 10 floors symbolise the Victorian gold rush of the 1850s and the red strip symbolises the bloodshed during the rebellion.

The building includes some sustainable features such as natural ventilation, high performance glass windows with maximised area thus minimising heat gain and loss for reducing energy consumption and natural lighting, etc.

After 10 floors of podium, the Eureka Tower has a diamond shaped plan section with a central core shear wall, perimeter columns, two shear walls on north-south direction and reinforced concrete outriggers linking the central core shear wall to the perimeter columns and 2 shear walls (Figure 4.25). Unlike most of the buildings with an outriggered frame structural system having outriggers located at one or more levels, the Eureka Tower has continuous outriggers, with 30 cm thickness, almost throughout the height of the building. To provide adequate stiffness, east-west and north-south outrigger shear walls are located in between the 11th to 65th floors and 11th to 89th floors respectively. The floor slabs are composed of post-tensioned reinforced concrete flat beams with 45 cm deep and 150 cm wide.

A wind tunnel test was carried out on a 1/400 scale model of the Eureka Tower and to limit the top drift, a liquid mass damper is used at the top of the building (between the 90th and 91st floors).

High-strength concrete is used to maximise the net usable area and to minimise the dimensions of structural elements. For the first 15 storeys, the strength of the concrete is 80 MPa in core shear walls with 75 cm thickness and 100 MPa in perimeter columns. The strength of the concrete gradually changes towards top of the building, decreasing to 40 MPa and 60 MPa for the core shear walls and perimeter columns respectively.

Bibliography

Binder, G., *101 of the World's Tallest Buildings,* pp. 204–205, Victoria: Images Publishing Group, 2006.

Cement and Concrete Association of Australia, *Mix Magazine*, High Strength Concrete: Recent Developments with 100 MPa at The Eureka Tower, No. 12, pp. 2–5, July 2003.

Dean, B., Martin, O., Emery, D. and Chancellor, P., *Tall Building Design Innovations in Australia,* pp. 393–399, CTBUH 6th World Congress 2001, Spon Press, Taylor & Francis Group, London and New York, 2001.

Eureka Tower, http://ctbuh.org, accessed May 2012.

Eureka Tower, http://www.emporis.com, accessed May 2012.

Freyssinet Magazine, Two Prestressed Skyscrapers, No. 213, p. 16, January/April 2002.

Martin, O. and Dean, B., *Advances in the Structural Design of High-Rise Residential Buildings in Australia,* CTBUH 7th World Congress 2005, New York, 2005.

The Architectural Review, Eureka Tower, Melbourne, Australia Fender Katsalidis Architects, No. 1340, October 2008.

World Trade Center Twin Towers

OFFICIAL NAME: World Trade Center Twin Towers (WTC I–II)
LOCATION: New York, USA
BUILDING FUNCTION: Office
ARCHITECTURAL HEIGHT: 415/417 m
NUMBER OF STOREYS: 110
STATUS: Demolished after collapse following fire and airplane impact (2001)
COMPLETION: 1972 (WTC I), 1973 (WTC II)
ARCHITECT: Minoru Yamazaki & Associates; Emery Roth & Sons
STRUCTURAL ENGINEER: Leslie E. Robertson Associates
STRUCTURAL SYSTEM: Framed-tube system/steel

World Trade Center Twin Towers: architectural and structural information

The 110-storey, 417/415 m high World Trade Center Twin Towers in New York (USA) were designed by Minoru Yamazaki & Associates and Emery Roth & Sons. They were steel buildings with a framed-tube system. The WTC I gained the title of "the world's tallest building" in 1972 and the WTC II did so in 1973. The buildings' structure was integrated with the architectural design and became prominent.

The World Trade Center Twin Towers (WTC I–II) were described by Ada Louise Huxtable as "Experimental architecture that goes beyond contemporary standards to explore a more evocative world for a broader, richer and more ornamental contemporary architecture. This large and important group of structures will be unlike anything that New York has ever seen before" (quoted in Harris, 2001, p. 38). Taking inspiration from the IBM Building (Seattle, 1964), the design for the buildings was chosen in a competition held in 1962 in which well-known architects such as Philip Johnson and I.M. Pei participated.

The World Trade Center consisted of 7 buildings in total and its twin towers were similar but not identical. WTC I (north tower) was 417 m high with a 110 m antenna and WTC II (south tower) was 415 m high. Both buildings had approximately 63×63 m square cross-sections with 2.1 m corner cuts. The floor-to-floor height was 3.66 m.

The framed-tube, consisting of closely spaced perimeter/exterior columns connected with deep spandrel beams, was designed to support the entire lateral load and 40 per cent of the vertical load, while the central core was designed to support 60 per cent of the vertical load and none of the lateral load (Figure 4.26).

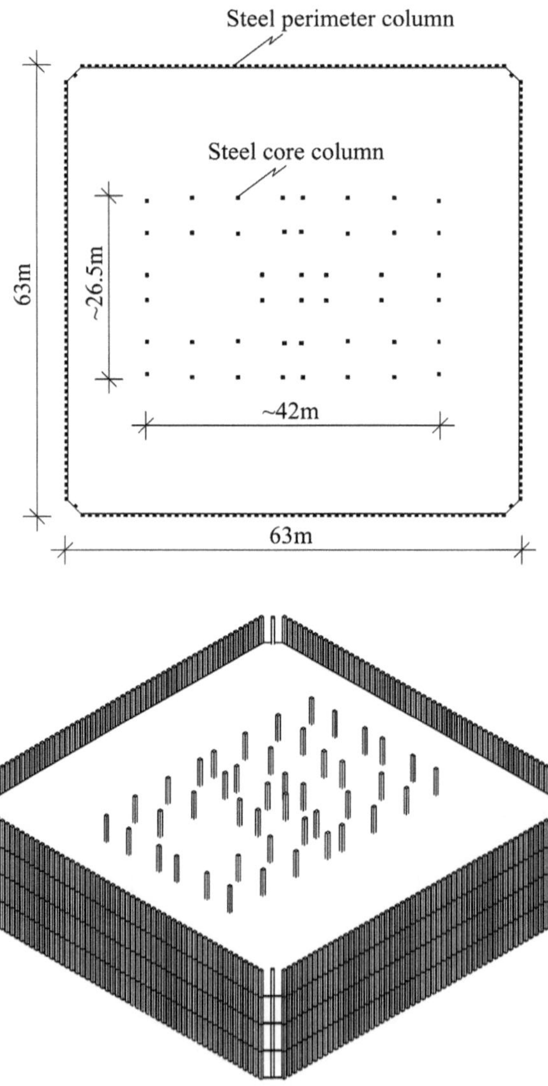

FIGURE 4.26 WTC I–II plan and structural axonometric

Closely spaced steel perimeter columns, with a box section of 36×34 cm varying by 10 to 0.6 cm in plate thickness from bottom to top, were connected at each floor with deep steel spandrel beams 132 cm in depth and a plate thickness of 3.6 to 0.9 cm from bottom to top. The columns were spaced at 102 cm centres, with a 66 cm clear span (Figure 4.27).

In order to create a wide and inviting space at the building entrances, 59 closely spaced perimeter columns were located on every side of the upper floors and from the 8th floor, three columns that merged into one descended to the entrance (Figure 4.28).

The 26×42 m core in the centre of the buildings consisted of columns on the lower storeys with 40×91 cm box-sections. On the upper storeys these columns changed to

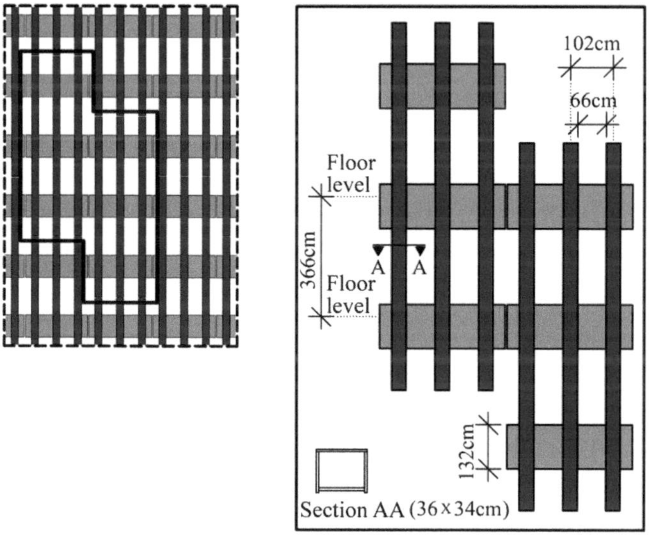

FIGURE 4.27 WTC I–II facade layout

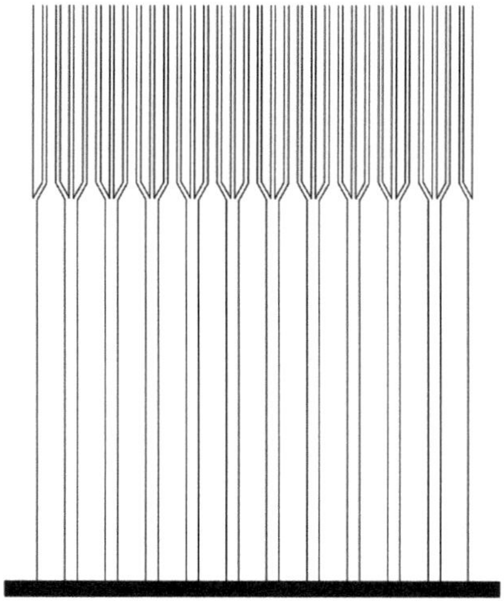

FIGURE 4.28 WTC I–II branching columns

"H" cross-sections. The core columns, varying in number between 44 and 47, were supported by steel braces at the mechanical equipment floors and the atrium at the ground floor.

Outriggers were used at the roof to contribute to the framed-tube system's stiffness against wind induced lateral loads (Figure 4.29). The outriggers connected the core and the perimeter columns in both directions.

The composite floor slab system of WTC I–II (Figure 4.30) outside the core consisted of a row of main double trusses spaced 203 cm apart with a depth of 74 cm spanning spaces of 18.3 m and 10.7 m, and, perpendicular to these, a 10 cm thick reinforced concrete layer on 3.8 cm thick trapezoidal metal deck resting on transverse (secondary) trusses spaced 406 cm apart. The floor slabs were 13.8 cm thick outside the core and 16.5 cm in the core.

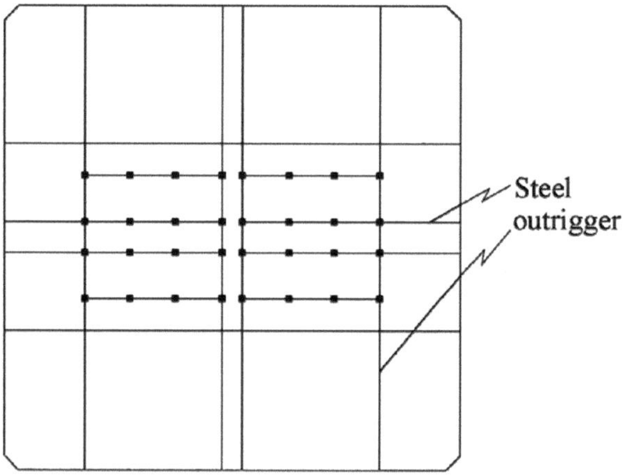

FIGURE 4.29 WTC I–II outrigger location at the upper storeys

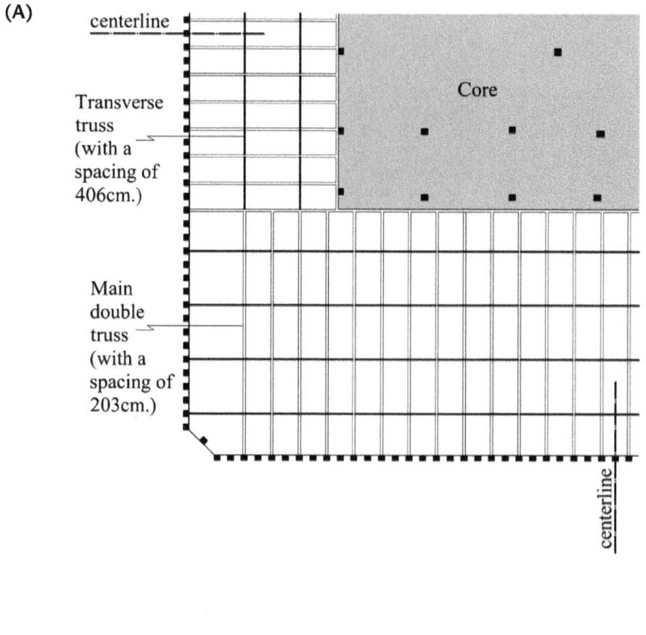

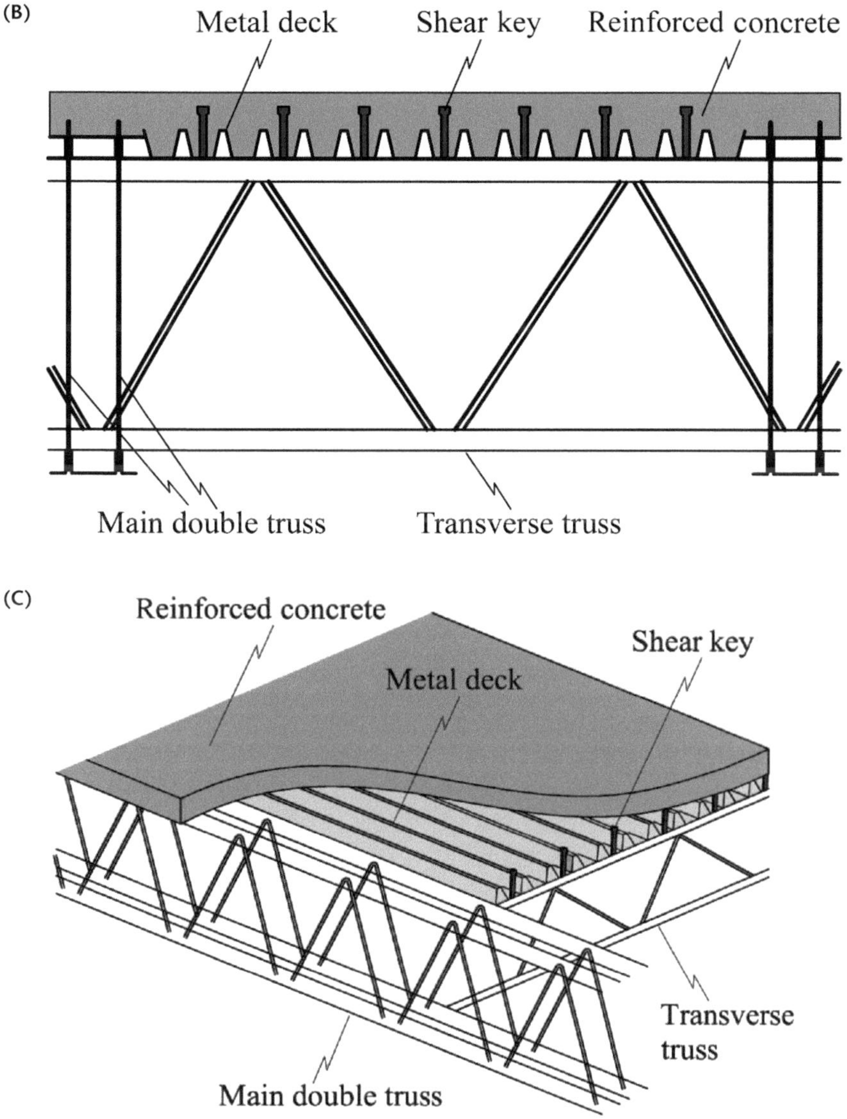

FIGURE 4.30 WTC I–II slab system: (a) plan, (b) section, (c) axonometric

Bibliography

Corley, W.G., *Structural Investigation of the New York World Trade Center Collapse*, Proceedings of the CIB-CTBUH International Conference on Tall Buildings, Malaysia, 2003.

Federal Emergency Management Agency (2001) Federal Insurance and Mitigation Administration, *World Trade Center Building Performance Study*, FEMA Report, 2001.

Fukuda, R., Fukuzawa, E., Koshika, N., Morikawa, H., Yoshikai, S., Makibe, K. and Mihara, Y., *Analyses of Structural Responses and Damages of World Trade Center Tower 2 Due to Impact of Aircraft Crashed*, Proceedings of the CIB-CTBUH International Conference on Tall Buildings, Malaysia, 2003.

Godfrey, B., *Multi-Storey Buildings in Steel*, Second Edition, Collins, London, England, pp. 162–164, 1985.
Harris, B., *The World Trade Center: A Tribute*, Running Press Book Publishers, The United States, 2001.
Lew, H.S. and Sadek, F., *WTC Towers: Innovative Design Features and Structural Modeling*, CTBUH Seoul Conference, pp. 1150–1153, Seoul, 2004.
Taranath, B., *Wind and Earthquake Resistant Buildings: Structural Analysis and Design*, A Series of Reference Books and Textbooks (Editor: Michael D. Meyer), Department of Civil and Environmental Engineering, Georgia Institute of Technology, Atlanta, Georgia, 2005.
World Trade Center Twin Towers, http://www.ctbuh.org, accessed May 2012.
World Trade Center Twin Towers, http://www.emporis.com, accessed May 2012.

John Hancock Center

OFFICIAL NAME: John Hancock Center
LOCATION: Chicago, ABD
BUILDING FUNCTION: Mixed-use (residential, office)
ARCHITECTURAL HEIGHT: 344 m
NUMBER OF STOREYS: 100
STATUS: Completed
COMPLETION: 1969
ARCHITECT: Bruce Graham (Skidmore, Owings & Merrill – SOM)
STRUCTURAL ENGINEER: Fazlur Rahman Khan (SOM)
STRUCTURAL SYSTEM: Trussed-tube system/steel

(photos courtesy of Marshall Gerometta/CTBUH)

John Hancock Center: architectural and structural information

The 100-storey, 344 m high John Hancock Center in Chicago (USA) was designed by Bruce Graham (SOM) and Fazlur Rahman Khan (SOM). It is a steel building with a trussed-tube system. The innovation of utilising the trussed-tube (braced-tube) system to carry the lateral loads was first used by the structural engineer Fazlur Rahman Khan in the structural design of the John Hancock Center. The expressive trussed-tube structure of the building was integrated with the architectural design and became prominent. The John Hancock Center represents an icon of Chicago and symbol of structural expression in architecture. Fazlur Rahman Khan commented that "The social and visual impact of buildings is really my motivation for searching out new structural systems and to get the right visual impact, a building's natural strength should be expressed" (Mufti and Bakht, 2002). The John Hancock Center was granted the "Distinguished Building Award" in 1970 by AIA (American Institute of Architects).

The John Hancock Center in Chicago has an important place among tall buildings. The exterior mega cross braces of the trussed-tube system, which can be seen on the tapering facade, emphasise the architectural aesthetic of the structural system and are the most striking design feature. The building is an architectural icon, with a sloping form (truncated pyramid), narrowing as it rises and symbolises the integration of structural expression with the architecture. Thanks to its original design, in 1970 it won the "Honor Award of the AIA Chicago Chapter" in the architectural and structural engineering section and in 1999 it won the "25-Year Architectural Excellence Award of the AIA" for its preservation of architectural excellence and authenticity for more than 25 years.

Bruce Graham (the architect) and John Hancock Insurance (the owner) are said to have wanted to remove the exterior X-braces on the upper 10 storeys of the building because they blocked the view; however, Fazlur Rahman Khan (the structural engineer), aiming to integrate the architecture with the structural system, asserted that the structure would give great aesthetic value to the building and convinced the architect and owner to repeat the exterior braces, consisting of trusses, continuously throughout the height of the building. Thanks to Fazlur Rahman Khan's design, the continuous X-braces expressed on the facade with their reduced dimensions and cross-sectional areas towards the top of the building contribute to the design approach of the tapered form of the building, which narrows as it rises, creating the impression that the building is taller than it is in reality.

Mega X-braces on the facade, designed as truss elements with 45° angles between them, support a large part of the wind induced lateral loads that are converted to an axial load and also support a part of the vertical load, ensuring the system's pure tubular behaviour.

The John Hancock Center's truncated pyramid form and thus reduction of the floor plan area, has several advantages. The placing of office areas, which need large spaces and long lease spans, on the lower floors and residential areas, which need smaller spaces and shorter lease spans, on the upper floors, provides a continuous relationship between form and function. With the tapering form by the inward slope of the facade, the surface area affected by the wind on the higher levels of the building is reduced, as is the wind intensity and thus the excess wind pressure/load (Section 6.1). In calculating the aspect ratio (the ratio of the structural height of a building to the narrowest structural width on the floor plan), because the widest dimensions of the floor plan of the building were taken as a basis, the wide ground floor layout improves this ratio and thus the slenderness and flexibility of the building. Due to the widening of the building facade as it descends towards the ground, the potential for the view from the higher floors to cause dizziness and vertigo is reduced.

The ground floor of the John Hancock Center has a rectangular cross-section approximately 50×80 m with a distance of 17.5 m of lease span – between the core and the perimeter – on the long face and the top floor has a rectangular cross-section approximately 30×49 m with a distance of 7.5 m of lease span on the short face (Figure 4.31).

In the John Hancock Center, perimeter columns are spaced at approximately 12 m centres on the long faces and at approximately 7.5 m centres on the short faces of the building, with exterior X-braces on the facade and spandrel beams connecting all these together (Figure 4.32).

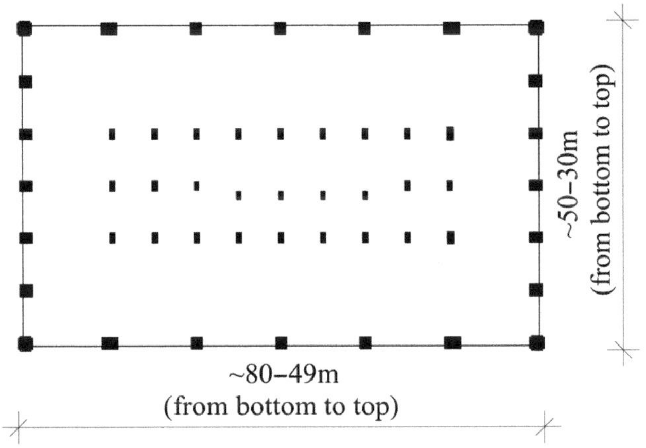

FIGURE 4.31 John Hancock Center plan

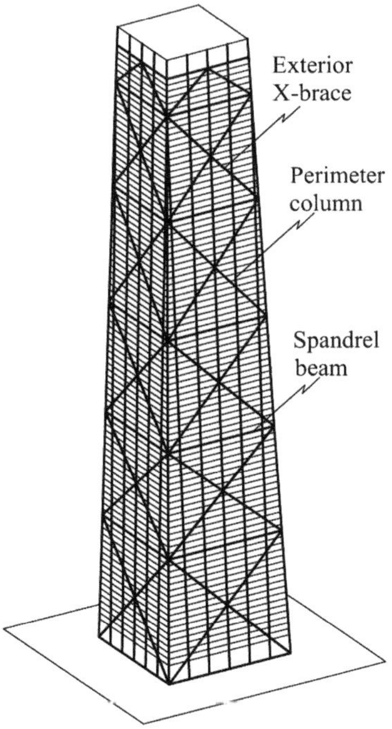

FIGURE 4.32 John Hancock Center structural axonometric

Bibliography

Iyengar, H., Reflections on the Hancock Concept, *CTBUH Review*, Vol. 1, p. 1, May 2000.

John Hancock Center, http://www.ctbuh.org, accessed May 2012.

John Hancock Center, http://www.emporis.com, accessed May 2012.

Fazlur Rahman Khan – John Hancock Center, http://khan.princeton.edu, accessed May 2012.

Mufti, A.A. and Bakht, B., Fazlur Khan (1929–1982): Reflections on his life and works, *Canadian Journal of Civil Engineering*, Vol. 29, pp. 238–245, 2002.

Taranath, B., *Steel, Concrete & Composite Design of Tall Buildings*, New York: McGraw-Hill Book Company, 1998.

Weingardt, R.G., Great Achievements – Notable Structural Engineers: Fazlur Rahman Khan, The Einstein of Structural Engineering, *Structure Magazine*, February 2011.

Zaknic, I., Smith, M. and Rice, D., *100 of The World's Tallest Buildings*, Council on Tall Buildings and Urban Habitat Committee, New York: McGraw-Hill Book Company, 1998.

CCTV Headquarters

OFFICIAL NAME: CCTV Headquarters
LOCATION: Beijing, China
BUILDING FUNCTION: Office
ARCHITECTURAL HEIGHT: 234 m
NUMBER OF STOREYS: 49
STATUS: Completed
COMPLETION: 2011
ARCHITECT: Office for Metropolitan Architecture
STRUCTURAL ENGINEER: East China Architectural Design and Research; Ove Arup & Partners
STRUCTURAL SYSTEM: Trussed-tube system/composite

(photos courtesy of M. Bunyamin Bilir)

CCTV Headquarters: architectural and structural information

The 49-storey, 234 m high CCTV Headquarters in Beijing (China) was designed by Office for Metropolitan Architecture. It is a composite building with a trussed-tube system and can also be named as a loop-tube system.

The CCTV Headquarters, where units providing all television programme production, broadcasting and management services are based, has a looped form consisting of an "L" cross-section at the base upon which are 2 sloping towers from which extend cantilevers with an "L" cross-section. It is the first and only building in which a trussed-loop-tube system has been used.

At the ground floor the building has a 9-storey "L" cross-section base and the two 36-storey prismatic sections rise vertically in both directions at a 6 degree slope, above which the 9–13 storey section also has an "L" cross-section form, like the base. In the design, greater importance was given to the form than the height.

The tube system was considered to be the most appropriate structural system for the form of the CCTV Headquarters. Tubular behaviour was obtained with a trussed-tube system consisting of steel and composite columns, steel beams and steel braces. The arrangement of the braces on the perimeter of the looped form building was finalised as a result of structural analysis (Figure 4.33). The exterior braces are spaced twice as closely in areas that have high interior forces and are reduced by half in areas with low interior forces.

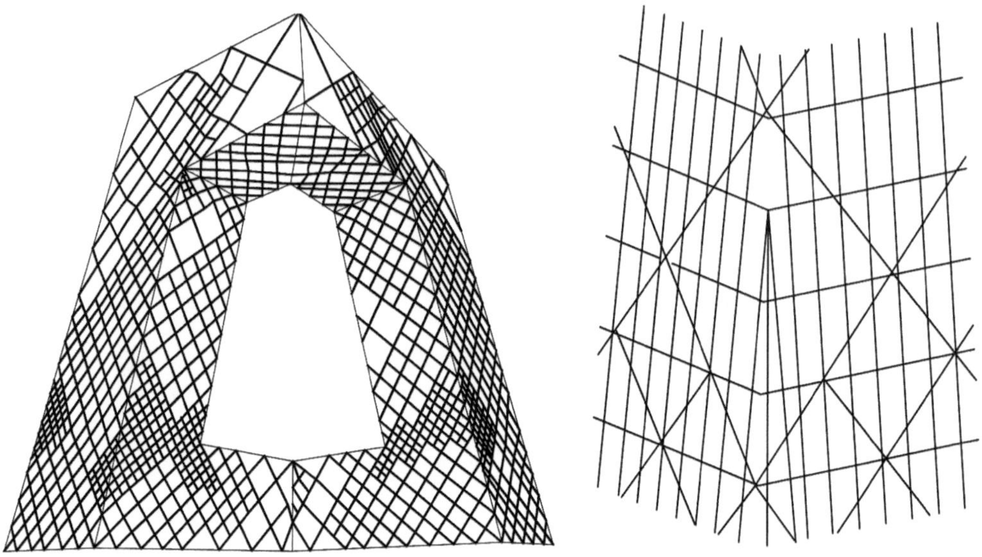

FIGURE 4.33 CCTV Headquarters steel facade grid (Koolhaas and Scheeren, 2005)

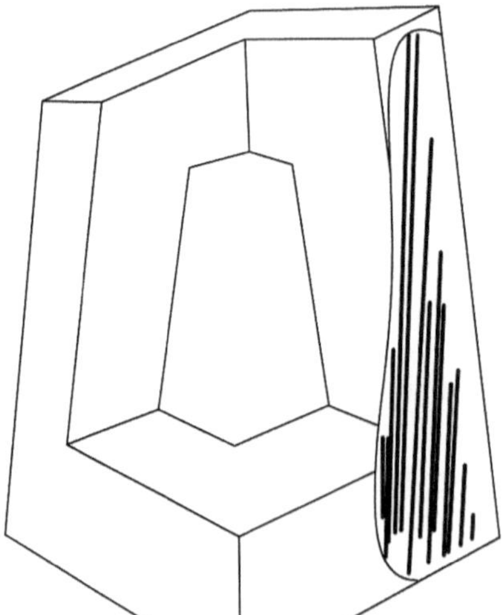

FIGURE 4.34 CCTV Headquarters discontinuous columns (Carroll et al., 2008)

Because some of the interior columns in the sloping towers and all of the interior columns in the cantilever section, do not continue throughout the height of the building from top to bottom, owing to the form of the building, 2-storey-deep transfer beams were used in order to transfer the column loads (Figure 4.34). The transfer beams are situated on the bottom two storeys of the cantilevered section and on the mechanical equipment floors in the sloping towers.

Bibliography

Carroll, C., Cross, P., Duan, X., Gibbons, C., Ho, G., Kwok, M., Lawson, R., Lee, A., Li, R., Luong, A., McGowan, R. and Pope, C., Case Study: CCTV Building, *CTBUH Journal*, Issue 3, 2008.

Carroll, C., Gibbons, C., Ho, G., Kwok, M., Lawson, R., Lee, A., Li, R., Luong, A., McGowan, R. and Pope, C., CCTV Headquarters, Beijing, China: Building the Structure, *The Arup Journal*, No. 2, 2008.

Carroll, C., Cross, P., Duan, X., Gibbons, C., Ho, G., Kwok, M., Lawson, R., Lee, A., Luong, A., McGowan, R. and Pope, C., CCTV Headquarters, Beijing, China: Structural Engineering Design and Approvals, *The Arup Journal*, No. 2, 2005.

CCTV Headquarters, http://www.ctbuh.org, accessed May 2012.

CCTV Headquarters, http://www.emporis.com, accessed May 2012.

Green, D., Lau, B., Lyall, I., Pritchard, D., Pullen, J. and Ryan, C., CCTV Headquarters, Beijing, China: Services Engineering Design, *The Arup Journal*, No. 3, 2005.

Koolhaas, R. and Scheeren, O., *CCTV by OMA*, A+U Co., Ltd., Japan, July 2005.

Willis Tower

OFFICIAL NAME: Willis Tower (formerly Sears Tower)
LOCATION: Chicago, USA
BUILDING FUNCTION: Office
ARHITECTURAL HEIGHT: 442 m
NUMBER OF STOREYS: 108
STATUS: Completed
COMPLETION: 1974
ARCHITECT: Bruce Graham (Skidmore, Owings & Merrill – SOM)
STRUCTURAL ENGINEER: Fazlur Rahman Khan (SOM)
STRUCTURAL SYSTEM: Bundled-tube system/steel

(courtesy of Antony Wood/CTBUH)

Willis Tower: architectural and structural information

The 108-storey, 442 m high Willis Tower in Chicago (USA) was designed by Bruce Graham (SOM) and Fazlur Rahman Khan (SOM). It is a steel building with a bundled-tube system. The innovation of utilising the bundled-tube system to carry the lateral loads was first used by the structural engineer Fazlur Rahman Khan. The Willis Tower gained the title of "the world's tallest building" in 1974 and held it for 24 years (1974–1998). It was the winner of the "Distinguished Building Award" in 1976 by

AIA (American Institute of Architects). While the original name of the building was the Sears Tower, it was renamed the Willis Tower in 2010. The first use of steel in a bundled tube system was in the Willis Tower. The structure of the building was integrated with the architectural design and became prominent. The expressive tubular structure of the building in the form of bundled-tube was integrated with the architectural design and became prominent. The Willis Tower and the John Hancock Center are considered as Fazlur Rahman Khan's masterpieces. Like the John Hancock Center, the Willis Tower represents an icon of Chicago and a symbol of structural expression in architecture.

In the design of the Willis Tower, a planning approach that the structural elements in the interior space did not obstruct the architecture was used, in an effort to create a space with less structural material and thus less cost.

The Willis Tower is an example of a building using setbacks on the building facade in order to reduce the effect of the wind by breaking up its flow (Section 6.1).

The building has a 68.7×68.7 m square cross-section at the ground level and is formed from 9 rectangular tubes with 22.9×22.9 m square cross-sections. The form of the Willis Tower begins with 9 tubes at the base, with 2 tubes ending at the 50th floor,

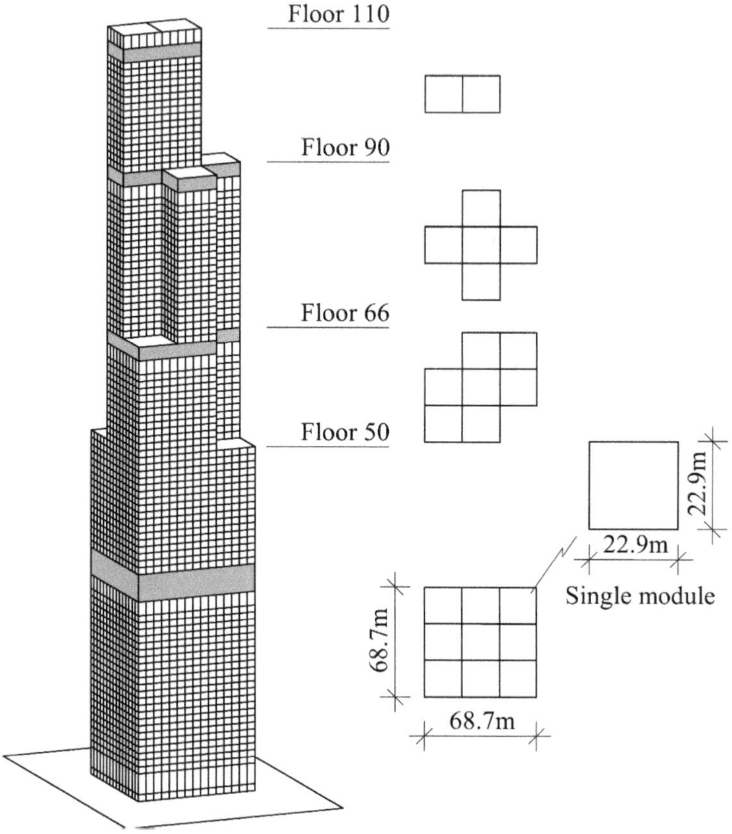

FIGURE 4.35 Willis Tower axonometric and schematic plans

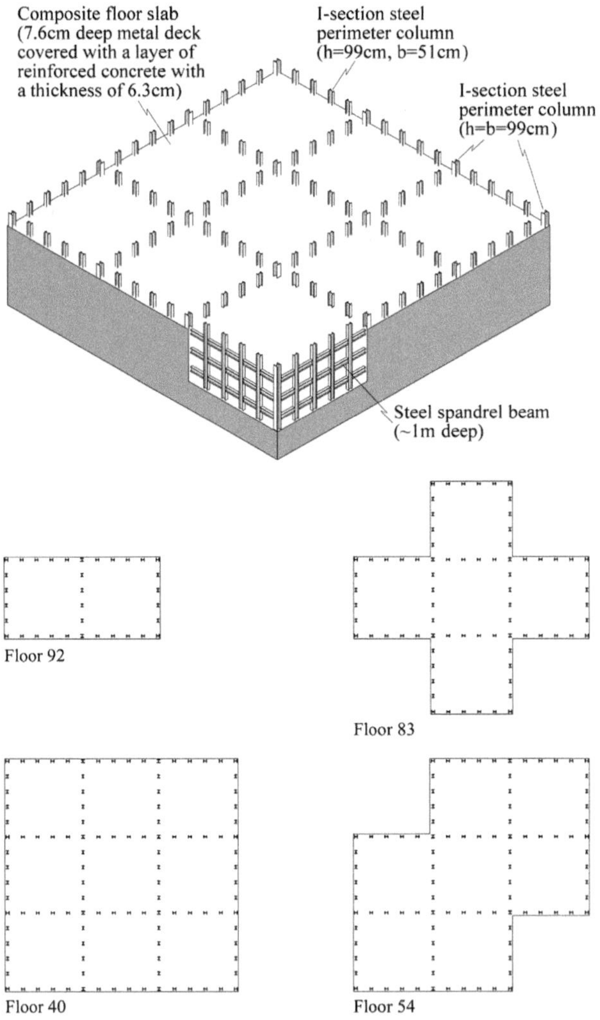

FIGURE 4.36 Willis Tower structural axonometric and plans

2 tubes ending at the 66th floor, 3 tubes ending at the 90th floor and 2 tubes ending at the top (Figure 4.35–4.36). The floor-to-floor height is 3.92 m.

In the Willis Tower, columns are spaced at 4.6 m centres within the building and on the facade and connected at each floor with deep spandrel beams (Figure 4.36). The columns and beams consist of steel elements with approximately 1 m deep I-section, reducing in their plate thickness and the length of their flanges throughout the height of the building. These elements are formed of modular tube units which are able to reach the desired height without compromising structural strength.

The Willis Tower's composite floor slabs are composed of 7.6 cm deep trapezoidal metal deck covered with a layer of reinforced concrete with a thickness of 6.3 cm and supported by approximately 0.9 m deep steel trusses spanning 23 m and spaced every 4.6 m (Figure 4.36).

Bibliography

Bennett, D., *Skyscrapers: Form & Function*, Simon & Schuster Ltd., New York, 1995.

Fazlur Rahman Khan – Sears Tower, http://khan.princeton.edu, accessed May 2012.

Taranath, B., *Steel, Concrete & Composite Design of Tall Buildings*, New York: McGraw-Hill Book Company, 1998.

Taranath, B., *Wind and Earthquake Resistant Buildings: Structural Analysis and Design*, A Series of Reference Books and Textbooks (Editor: Michael D. Meyer), Department of Civil and Environmental Engineering, Georgia Institute of Technology, Atlanta, Georgia, 2005.

Willis Tower, http://www.ctbuh.org, accessed May 2012.

Willis Tower, http://www.emporis.com, accessed May 2012.

Willis Tower, http://www.willistower.com, accessed May 2012.

Zaknic, I., Smith, M. and Rice, D., *100 of The World's Tallest Buildings*, Council on Tall Buildings and Urban Habitat Committee, New York: McGraw-Hill Book Company, 1998.

Commerzbank Tower

OFFICIAL NAME: Commerzbank Tower
LOCATION: Frankfurt, Germany
BUILDING FUNCTION: Office
ARCHITECTURAL HEIGHT: 259 m
NUMBER OF STOREYS: 56
STATUS: Completed
COMPLETION: 1997
ARCHITECT: Foster + Partners
STRUCTURAL ENGINEER: Ove Arup & Partners; Krebs und Kiefer
STRUCTURAL SYSTEM: Mega column system/Composite

Commerzbank Tower: architectural and structural information

The 56-storey, 259 m high Commerzbank Tower in Frankfurt (Germany) was designed by Foster + Partners. The Commerzbank Tower gained the title of "the tallest building in Europe" in 1997. It won the "Green Building Award of the City of Frankfurt" in 2009 in recognition of the building's pioneering role in environmentally-friendly and energy-saving architecture. Other prestigious awards include "RIBA Architecture Award", "Bund Deutscher Architekten – Martin-Elsaesser-Plakette Award" and "British Construction Industry Award".

Norman Foster (the architect), indicated that two important factors were central to the design of the Commerzbank Tower: i) the transparency of the building to light and to views and ii) the incorporation of nature. These two unique design features were attained by the innovative structural design of the building. The structural and environmental innovations were the major success factors of the design of the building.

The environmentally conscious Commerzbank Tower has an equilateral triangular plan with gently rounded corners and slightly curved sides each measuring at about 60 m (Figure 4.37). As a result of this plan scheme, the building performs better against wind pressure compared to a building having a rectangular plan.

The building's main design feature is the central triangular atrium and its relationship with the corner cores. This full-height central atrium is supported by triangular steel columns at the corners which vary 140 cm to 60 cm from bottom to top. The central atrium is surrounded by landscaped sky gardens, which spiral up the building to form the visual and social focus of office floors (Figure 4.38). Three different style gardens have been proposed according to the direction: i) east-facing sky-gardens have "Oriental Style", ii) south-facing sky-gardens have "Mediterranean Style", iii) west-facing sky-gardens have "North American Style". These landscaped spaces can be used for recreation; therefore they improve the working environment in an immeasurable way. On any one level, one side of the building opens to sky gardens, each four storeys high (14.02 metres) and linked to the central triangular atrium. While one side of the building opens to a sky garden, the other two sides office spaces for a working group of about 40 people.

The glass curtain wall enclosed sky gardens also improve the environmental conditions inside the building, bringing daylight and fresh air into the central atrium, which acts as a natural ventilation chimney for the inward-facing offices. Hence, all offices can gather direct sunlight and fresh natural air due to the availability of sky

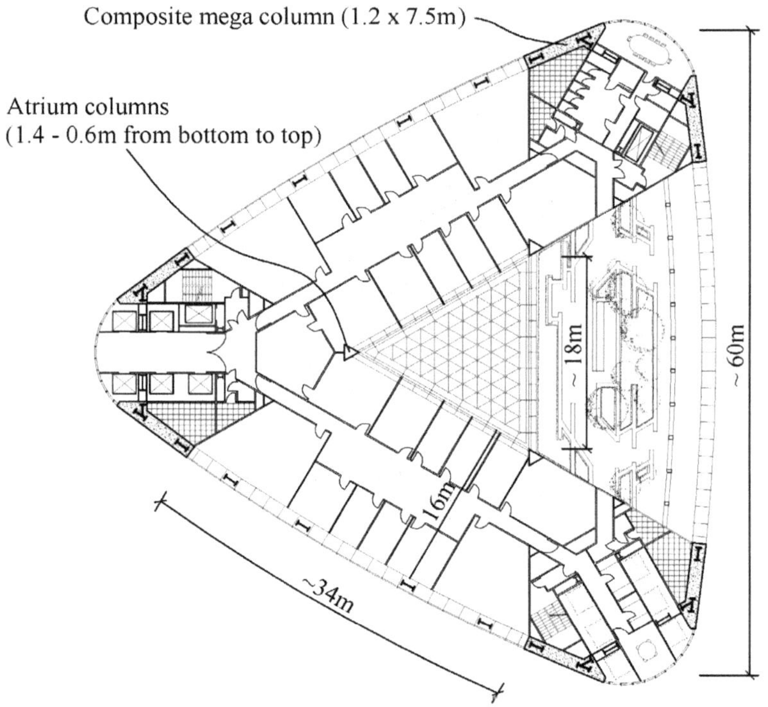

FIGURE 4.37 Commerzbank Tower plan

FIGURE 4.38 The relationship of the central atrium with the spiralling sky gardens

gardens and the central atrium. The day-lit spaces increase the energy efficiency due to reduced artificial lighting.

The façades of the exterior-facing office spaces are treated by an aluminum curtain wall which acts as an air cavity from bottom to top of building. The inside layer of the curtain wall system consists of a double-glazed window which acts as the main envelope between inside and out. The outside layer, on the other hand, is a fixed glass and acts as the second layer. The air enters the ventilated cavity between these two layers at sill level and is expelled through a slot at the top of the window. Unlike the traditional glare control mechanisms, the blinds that reduce glare in the office as well as solar heat gain are located inside the cavity. As a result, these blinds stop the heat before it enters the building. The naturally ventilated cavity with user controlled blinds improves the thermal insulating properties of the building. Post-occupancy studies have shown that the tower actually consumes about 20 per cent less energy than predicted. This is largely because the building users have extended the period of natural ventilation up to about 85 per cent of the year, as opposed to the 60 per cent designed for.

As a result of the central atrium, the core functions (e.g. vertical circulation) were pushed to the corners of the building. This plan scheme is quite different than a traditional deep-planned tower design with centralised core and identical floors resulting in insufficient natural ventilation and lighting. The corner located lifts offer dramatic views of the city and their movement animates the outside of the building.

The building's unique design had been made possible by a structural system which is composed of corner cores consisting composite mega columns (shear walls) coupled by steel link frames and steel Vierendeel frames coupling these cores (Figure 4.39). The two composite mega columns in corners consist of diagonally braced two vertical steel H-section profiles encased in reinforced concrete. Each core, having two mega columns with dimensions 1.2x7.5 m, is connected to the other with the 8-storey-deep and 34 m spanning Vierendeel frames along the outside of the building (Figure 4.39). Besides connecting the corner cores, these frames provide the structure to span sky gardens between the cores.

The two mega columns at each corner are connected to each other with steel link frames. The members of this link frame are passing between the lift shafts and they have a dimension of 1 m deep for vertical members and 1.1 m deep for beams (Figure 4.39).

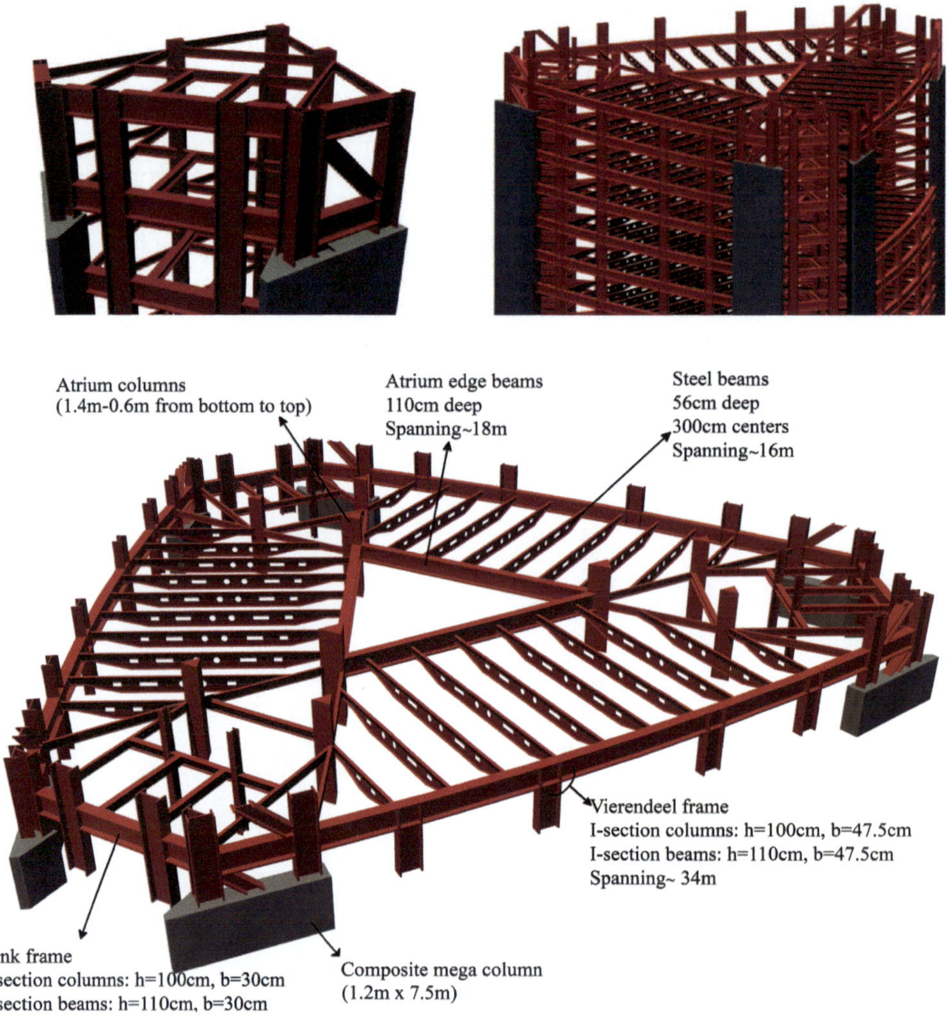

FIGURE 4.39 Commerzbank Tower structural axonometrics

Bibliography

Bailey, P., Bridges, H., Cross, P., Del Mese, G., Smith C., Walsh, S. and Wise, C., Commerzbank, Frankfurt, *The Arup Journal*, No 2, 1997.

Boake, T. M., Green Super Buildings, *Alternatives Journal*, November/December, 30/5, 2004.

Buchanan, P., Reinventing the Sky-Scrape, *A and U.* No 329, February, 30–67, 1998.

Foster+Partners, *Commerzbank Headquarters Frankfurt, Germany 1991–1997,* http://www.fosterandpartners.com/projects/commerzbank-headquarters, accessed June 2013.

5

THE EFFECT OF WIND ON TALL BUILDINGS

The history of skyscrapers, which epitomizes the twentieth century, begins in 1885 with the Home Insurance Building. Reaching a new and important point in 1931 with the Empire State Building and gathering speed in 1972 with the World Trade Center Twin Towers, this process is unfolding at an even greater rate today.

Since the weight of the structural system in the first skyscrapers made vertical forces more critical than lateral forces, wind loads were not considered important. In time, with developments and innovations in structural systems and the increase in the strength-to-weight ratio of the structural elements, the weight of buildings decreased and wind loads began to be important. Consequently, because the tall buildings being constructed today are lighter, more slender and more flexible than their predecessors, they are more prone to lateral drift with low damping, and wind-induced building sway has been transformed into one of the most important problems encountered by tall building designers, becoming a basic input to the design.

The wind loads affecting the building and the response of the building depend on the following factors:

- the characteristics of the wind
- the building size and geometry
- the stiffness of the building and the distribution of the building mass
- the inherent damping characteristics of the structural system and of the construction material, which dissipates wind-induced building sway
- the surrounding topography (issues with neighbouring buildings, etc.)
- the orientation (position) of the building with respect to the prevailing wind direction.

The determination of the acceptable sway limits of tall buildings is an important topic that has been extensively researched. Since wind speed and wind pressure increase according to the building height, wind loads become important as the building height increases. In general, structural design tends to be controlled by wind loads in tall and flexible buildings.

5.1 Wind-induced building motion

Wind-induced building motion can essentially be divided into three types (Figure 5.1):

- along-wind motion
- across-wind motion
- torsional motion.

In tall buildings, usually the across-wind motion and torsional motion are more critical than the along-wind motion. However, because of the complexity of the across-wind and torsional responses of a building, many existing building codes only suggest procedures to determine the along-wind response of the building.

5.1.1 Along-wind, across-wind and torsional motions

When an air mass moving in a particular direction makes contact with a building facade, it creates a force. This force, defined as the effect of the wind, increases as the wind speed or the surface area of the building facade exposed to the wind increases. Wind can affect more than one building face. When the wind force is perpendicular to the building face, the lateral drift is highest.

Building sway parallel to the direction of the wind is termed "along-wind motion" (Figure 5.1). This motion is induced by fluctuations in wind speed, and the variation in wind pressure between the windward (upstream) and leeward (downstream) faces (faces perpendicular to the wind direction) of the building.

Building sway perpendicular to the direction of the wind is termed "across-wind motion" (Figure 5.1). When the movement of the air mass is blocked by the building, because of its fluid behaviour it splits into two, passing both sides and the rear face of the building. Depending on the velocity of the wind, size and aspect ratio of the building, it will either smoothly flow through or circulate on the side and rear facades of the building. The compression of the streamlines around the sides of the building results in accelerated wind speeds in the shear layers, and thus vortices are formed by turbulent air flow (Figure 5.2).

Vortices are spiral flow formations generated by turbulence that create negative pressure in the across-wind direction while breaking away (shedding) from the surface of the building. They are shed alternately from either side (along the wind direction) of the building, following each other on opposite faces and interacting sequentially. As a result of this, because the forces developed on sides of the building cannot neutralise each other, in addition to the along-wind motion, across-wind motion occurs, which is usually more decisive and critical (Figure 5.3). Thus, while the windward face of the building is subject to positive wind pressure and the leeward face is subject to negative wind pressure (suction), the across-wind faces of the building are subject to alternating negative wind pressure. The rate at which the vortices are shed is a function of the building shape (Strouhal number), the building dimensions and the

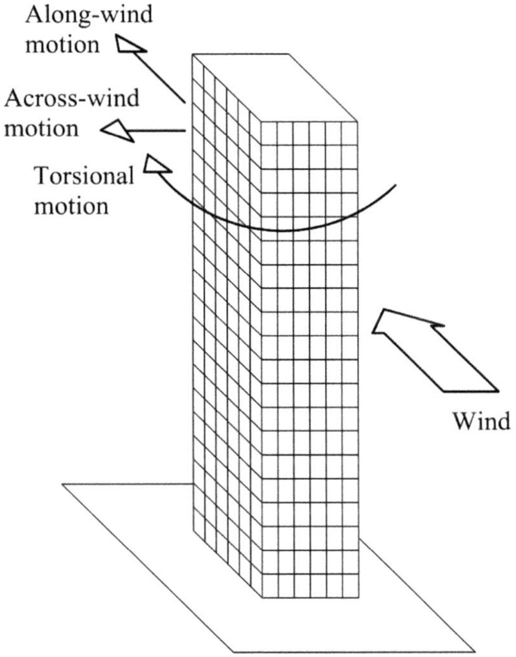

FIGURE 5.1 Tall building motions under the effect of wind

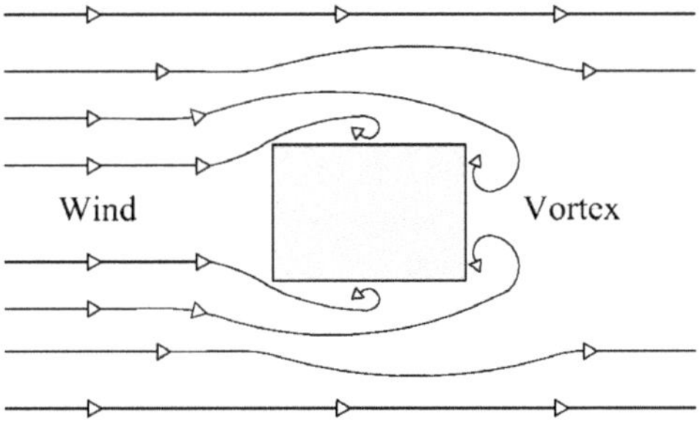

FIGURE 5.2 The formation of turbulent air flow

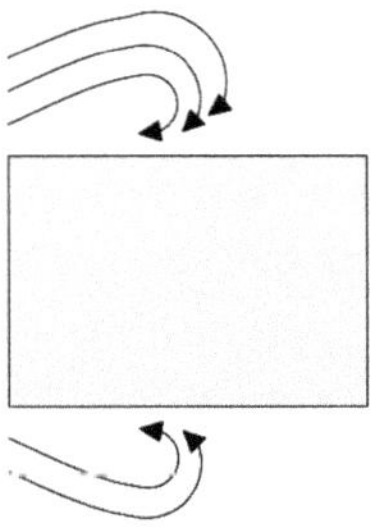

FIGURE 5.3 Vortices on along-wind sides of the building

wind speed. At higher wind speeds, the frequency of vortex shedding may approach the natural frequency of vibration of the building. When this occurs, the across-wind loads dominate over the along-wind loads, and the building motion can actually accentuate the vortex shedding leading to the phenomenon of negative aerodynamic damping. Aerodynamic architectural design and modifications (Chapter 6) reduce the strength of vortex shedding on tall buildings.

Even in buildings where the across-wind loads are smaller than the along-wind loads, the across-wind responses can still be critical for design as they will often still govern the serviceability accelerations. The serviceability accelerations are important in ensuring comfort for the occupants as perceptible motion can be disturbing if it is unexpected or occurs on a regular basis. Since in many tall and flexible buildings, across-wind building response is more critical than along-wind response (Holmes, 2001; Irwin, 2008; Kareem, 1985; Kwok, 1982), in general the concern to control across-wind response becomes a major design input. According to Gu and Quan (2004), "In the case of the Jin Mao Tower (Shanghai, 1999), the maximum acceleration in across-wind direction at design wind speed is about 1.2 times of that in the along-wind direction." Irwin (2008) states that "It is quite often the case that the highest overall wind loading on a tall slender building results from across-wind vortex excitation, which induces a large dynamic response."

As well as along-wind and across-wind responses, tall buildings may also experience torsional responses (Figure 5.1). These can occur if the shape of the building is asymmetric, if the structural system is asymmetric, or if the building is subjected to asymmetric flows.

The resultant wind force, acting perpendicular to the building face, passes through the geometric centre of the surface area of the building affected by the wind. On the other hand, the resultant reaction force passes through the stiffness centre of the building. When these two forces are not on the same axis, the resulting eccentricity creates torsional moments, resulting in floor torsion; thus, torsional motion occurs about the vertical axis of the building.

In general, most wind loading codes provide procedures for estimating the along-wind forces. Relatively few codes include procedures for across-wind and torsional responses, which by their nature are a lot less easily codified with accuracy. In the case of supertall buildings, along-wind, across-wind and torsional building responses, together with the dynamic effect of the wind, must be taken into account. In this context, dynamic calculation methods or wind tunnel tests are recommended for estimating the wind loads on such buildings.

5.2 Wind tunnel tests

As the height of buildings increases, wind loads become increasingly important for efficient and reliable design. While wind loads may influence the structural design of most tall buildings, for supertall buildings the minimisation of wind loads and responses can actually influence the architectural design. Shorter and/or less flexible buildings are generally treated by building codes as static structures, and wind load can be regarded as a static load on the building. For taller and/or more flexible structures the static load approach is insufficient, and the wind load on the building is treated as a dynamic load. Boundary-layer wind tunnel tests can be used to accurately determine the dynamic response of tall buildings to wind loading and excitation. In general, it is common to conduct wind tunnel tests for tall buildings having high aspect ratios. The boundary layer wind tunnel differs from aeronautical wind tunnels in that it models the inherent turbulence in the wind as well as the variation of wind speed with height above the ground surface, both of which must be modelled correctly in order to accurately predict the pressures and forces on buildings.

The sizes of the building and of the wind tunnel determine the scale of the model. In structural design, the wind load is determined by the results obtained from wind tunnel tests. In studies done to determine aerodynamic forces formed as a result of the interaction of the wind with tall buildings, the models used in wind tunnel tests are most commonly on scales of 1:300 and 1:500. Some recent examples are Chen et al., 2006 (1:500 scale Taipei 101 model) (Figure 5.4) and Weismantle et al., 2007 (1:500 scale Burj Khalifa model) (Figure 5.5). A guide to wind tunnel testing of high-rise buildings has recently been published by the CTBUH wind engineering working group which gives a more complete overview of the wind tunnel testing procedure (Irwin et al., 2013).

FIGURE 5.4 Taipei 101, wind tunnel test model (1:500 scale)
(credit for photo: RWDI)

FIGURE 5.5 Burj Khalifa, wind tunnel test model (1:500 scale) *(credit for photo: RWDI)*

6

DESIGN APPROACHES AGAINST WIND EXCITATION

At every step that tall building design takes toward the sky, today's architects and engineers encounter new difficulties. As the height of modern skyscrapers rises with developments in the field of structural system design and the use of high-strength materials, their weight and rigidity decrease, and their slenderness and flexibility – and thus their sensitivity to wind loads – increase. Wind loads, which cause large lateral drift, play a decisive role in the design of tall buildings and can be even more critical than earthquake loads. As a result, the wind loads and lateral drift to which tall buildings are subject have become an important problem (Figure 6.1).

Design approaches for controlling wind-induced building sway in tall buildings and protecting serviceability can be divided into three main groups, the "architectural design approach", the "structural design approach" and the "mechanical design approach" and their respective subgroups.

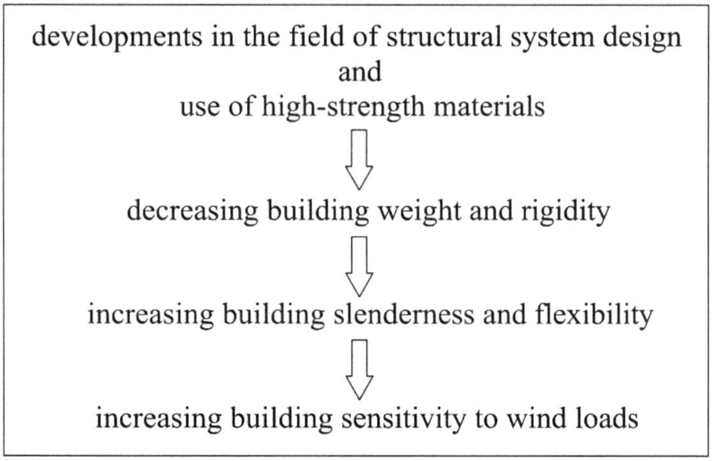

FIGURE 6.1 Tall buildings and the increasing building sensitivity to wind loads

6.1 Architectural design approach

6.1.1 Aerodynamic-based design

In tall and flexible buildings, aerodynamic behaviour generally becomes important. The wind-induced building response of tall buildings can be reduced by means of aerodynamic-based design and modifications that change the flow pattern around the building or break up the wind affecting the building face.

Aerodynamic-based design can be divided into two types, "aerodynamic architectural design" and "aerodynamic architectural modifications" and their subgroups.

Aerodynamic architectural design

Aerodynamic architectural design is realized by taking into consideration matters such as "building orientation (position)", "aerodynamic form", "plan variation" and "aerodynamic top" as part of the basic design. Aerodynamic architectural design plays an important role in reducing the effect of wind on tall buildings (Ali and Armstrong, 1995; Holmes, 2001; Irwin, 2009; Irwin et al., 2006; Irwin et al., 2008a; Irwin et al., 2008b; Kareem et al., 1999; Schueller, 1977; Scott et al., 2005). This reduction is generally in the region of 20–30 per cent, but can even exceed 50 per cent (Kim et al., 2008; Scott et al., 2005). These approaches are described below.

Building orientation (position)

Orienting (positioning) the building according to the prevailing wind direction is an effective design approach for reducing wind loads. A reduction of between 10–20 per cent of the across-wind building response can be obtained by rotating the building to within 10° of the wind direction (Scott et al., 2005). The effectiveness of this approach is dependent on both the wind climate at the project site and the shape of the building. In wind climates with very directional extreme winds and building shapes that are directionally sensitive this is more effective than, say, for a more regularly shaped building in a wind climate without strong directional characteristics.

Aerodynamic form

The use of aerodynamic building forms is an effective method of reducing the wind loads on buildings. In this context, cylindrical, elliptical, conical and twisted forms can be accepted among the efficient building forms.

Because cylindrical buildings (i.e., having circular or elliptical plan forms) have a smaller surface perpendicular to the wind direction, the wind pressure is less than in prismatic buildings. For buildings having circular plan form, the wind load is about 20 per cent less, compared with buildings having a rectangular plan form (Taranath, 2005). According to Davenport's study (1971) of models representing buildings of about 70 stories, the largest lateral drift value exhibited by a building with a circular plan under wind loads is approximately half of the lateral drift value exhibited by a building with a square plan. Buildings with elliptical plans also exhibit similar behaviour to buildings with circular plans. The architect of Le France Building in Paris stated that the wind load could be reduced by 27 per cent in buildings with an elliptical plan (Schueller, 1977).

Examples of buildings with aerodynamic forms include:

- Marina City Towers (Chicago, 1964) (Figure 6.2), with a cylindrical form
- Norman Foster's proposal of a conical form for the Millennium Tower (Tokyo, 1993, proposed) (Figure 6.3)
- The Bahrain World Trade Center (Manama, 2008) (Figure 4.4), with a sail-shaped form
- The Chicago Spire (Chicago, never completed) (Figure 6.4) and the Shanghai Tower (Shanghai, under construction), both of which have twisted forms.

Among these examples, aerodynamic form concerns played an important role in the architectural designs for the Millennium Tower and the Chicago Spire (Kareem et al., 1999; Tomasetti, 2007).

Due to the facade channels made possible by the twisted form of the design of the Chicago Spire, the effect of wind on the building is neatly blocked by breaking up the wind flow. Thus, wind-induced lateral loads are reduced. Thornton Tomasetti, the structural engineers of the building, reported that the design was the result of collaboration between the architects and the structural engineers, and that wind tunnel tests had confirmed the aerodynamic efficiency of the architectural form (Tomasetti, 2007).

FIGURE 6.2 Marina City Towers, Chicago, USA, 1964
(courtesy of Antony Wood/CTBUH)

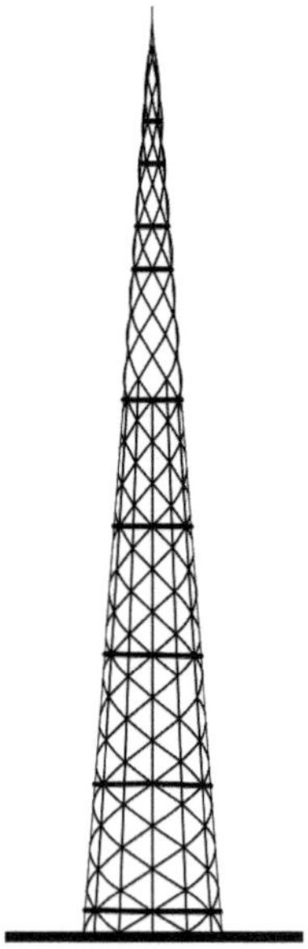

FIGURE 6.3 The Millennium Tower, Tokyo, Japan, 1993, proposed

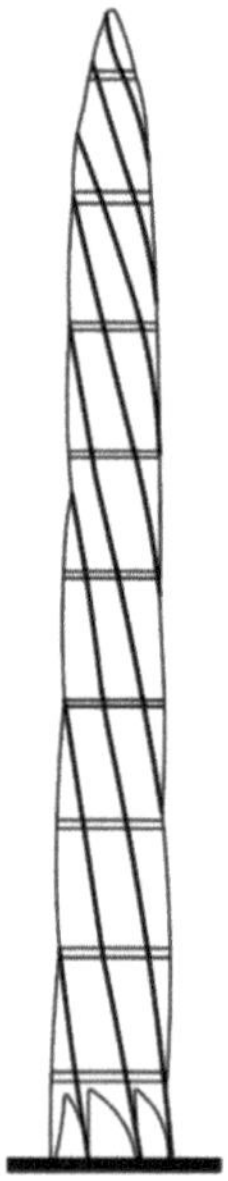

FIGURE 6.4 Chicago Spire, Chicago, USA, never completed

Plan variation

Varying the building plan results from the variation in plan dimensions or shape throughout the height of the building and can be achieved by:

a. reducing the plan area
b. changing the plan shape.

Plan variation by reducing the plan area toward the top of the building results in a reduction in the surface area affected by the wind at the upper levels of the building, which lessens the wind intensity and thus the excess pressure. The reduction in the plan area of the building as it rises can be in the form of:

- tapering
- setbacks.

Creating an inward-tapered facade (resulting in the building narrowing upward) or providing setbacks are effective methods for reducing the across-wind building response (Ali and Armstrong, 1995; Davenport, 1988; Irwin, 2008; Irwin, 2009; Irwin et al., 2008a; Irwin et al., 2008b; Kim et al., 2008; Schueller, 1977; Scott et al., 2005).

By designing tall and slender buildings in this way, lateral drift can be reduced by 10 to 50 per cent (Schueller, 1977). An analytical study by Khan (1972) has shown that, by creating a slope of 8 per cent in the facade of a 40-storey building, a 50 per cent reduction of the lateral drift in the upper stories can be obtained.

Examples of buildings with tapering include:

- John Hancock Center (Chicago, 1969) (Figure 6.5g)
- Chase Tower (Chicago, 1969) (Figure 6.6)
- Transamerica Pyramid (San Francisco, 1972) (Figure 6.7)
- Shanghai World Financial Center (Shanghai, 2008) (Figure 6.5c).

Buildings in which setbacks have been used to reduce the plan area:

- Petronas Twin Towers (Kuala Lumpur, 1998) (Figure 6.5d)
- Bank of China Tower (Hong Kong, 1989) (Figure 6.5f)
- Willis Tower (Chicago, 1974) (Figure 6.5e)
- Burj Khalifa (Dubai, 2010) (Figure 6.5a).

Among these examples, aerodynamic form played an important role in the architectural design of the Burj Khalifa from the earliest stages of the design (Irwin and Baker, 2006).

The Taipei 101, completed in 2004 (Figure 6.5b), is an example of the use of both setbacks and tapering. However, since the facades are tapered outward, in the form of repetitive modules, setback formation does not cause a reduction in the plan area toward the top of the building.

Varying the plan by changing the plan shape at various levels throughout the height of the building causes a corresponding change in the vortex shedding effect, which disorients the across-wind vortices and breaks up their organization (Irwin, 2009).

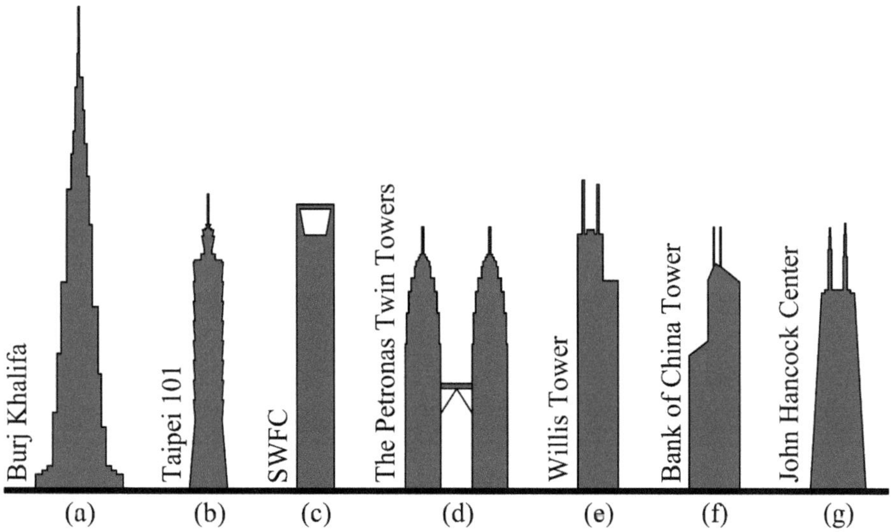

FIGURE 6.5 Some examples of buildings with tapering and setbacks

FIGURE 6.6 Chase Tower, Chicago, USA, 1969
(courtesy of Marshall Gerometta/CTBUH)

FIGURE 6.7 Transamerica Pyramid, San Francisco, USA, 1972
(courtesy of Niels Jakob Darger)

Aerodynamic top

The basis of the aerodynamic building top approach is the creation of an aerodynamic form near the top of the building that is part of the aerodynamic design of the building overall. These elements include approaches such as tapering the upper part of the building by progressively reducing the plan area and/or providing wind openings. Paying attention to the aerodynamics of the building top secures improvements not only in the along-wind, but also in the across-wind building response, by reducing the effect of wind-induced turbulence (vortex shedding forces) (Dutton and Isyumov, 1990; Ho, 2007; Irwin, 2009; Irwin et al., 2008a; Irwin et al., 2008b; Isyumov et al., 1992; Kareem et al., 1999). To reduce the across-wind response of the building, the optimum location for the along-wind openings is positioned between 80 per cent and 90 per cent of the building height (Kikitsu and Okada, 2003).

Examples of tall buildings with an aerodynamic top include:

- Taipei 101 (Taipei, 2004) (Figure 3.36)
- Jin Mao Building (Shanghai, 1999) (Figure 6.8)
- Two International Finance Centre (Hong Kong, 2003) (Figure 6.9)
- Petronas Twin Towers (Kuala Lumpur, 1998) (Figure 6.10)
- Central Plaza (Hong Kong, 1992)
- Shanghai World Financial Center (Shanghai, 2008) (Figure 6.11) (Ho, 2007; Kareem et al., 1999).

Among these examples, an aerodynamic top consisting of trapezoidal wind openings played an important role in the architectural design of the Shanghai World Financial Center (Kareem et al., 1999).

FIGURE 6.8 Jin Mao Building, aerodynamic top
(photo on right courtesy of Niels Jakob Darger)

FIGURE 6.9 Two International Finance Centre, aerodynamic top
(courtesy of Antony Wood/CTBUH)

FIGURE 6.10 The Petronas Twin Towers, aerodynamic top *(courtesy of Antony Wood/CTBUH)*

FIGURE 6.11 Shanghai World Financial Center, aerodynamic top *(courtesy of Niels Jakob Darger)*

Aerodynamic architectural modifications

Aerodynamic architectural modifications consist of corner modifications that do not significantly alter the existing architectural design. Modifications to corner geometry by means of recessed/notched, cut, slotted and rounded corners reduce the across-wind building response, as compared with an original building shape with sharp corners.

In a prismatic building, recessed (notched), cut, slotted and rounded corners (Figure 6.12) can reduce the along-wind and across-wind building response to an important degree (Gu and Quan, 2004; Irwin, 2009; Irwin et al., 2008a; Irwin et al., 2008b; Kawai, 1998; Kim et al., 2008; Kwok, 1995; Kwok and Bailey, 1987; Kwok et al., 1988; Scott et al., 2005). A chamfered (recessed/notched or cut) corner, which reduces the width of the building by 10 per cent compared with a sharp corner, reduces the along-wind building response by 40 per cent and the across-wind building response by 30 per cent (Holmes, 2001). Irwin (2009) terms "modified corners" as "softened corners" and states that "The corner softening should extend about 10 per cent of the building width in from the corner." However, corner modifications may cause adverse effects in serviceability and safety of the building (Kareem et al., 1999).

Rounded corners are the most effective type of corner modification (Gu and Quan, 2004). Approximating a circular plan form by increasing the corner roundness also reduces the wind loads affecting the building to an important degree (Gu and Quan, 2004; Kareem et al., 1999; Miyashita et al., 1995).

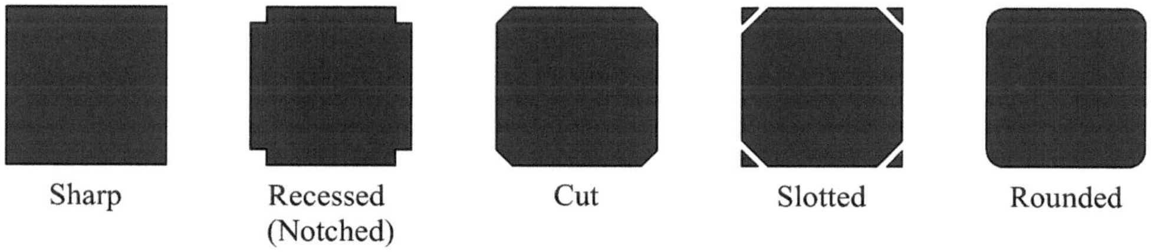

FIGURE 6.12 Modifications to corner geometry

FIGURE 6.13 Taipei 101, saw-tooth corners

FIGURE 6.14 Two International Finance Centre, saw-tooth corners

When comparing saw-tooth corners – which are a development of recessed corners – with sharp corners, in the view of Poon et al. (2004) they reduce the wind load affecting the building to an important degree. According to Irwin (2008, 2009) they cause approximately a 25 per cent reduction in the wind-induced base moment in the case of Taipei 101 (Figure 6.13).

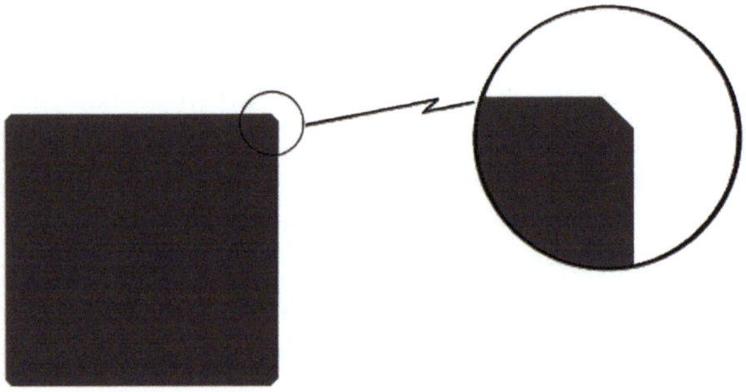

FIGURE 6.15 World Trade Center Twin Towers, cut corners

Examples include the use of saw-tooth (double-notch) corners in the Two International Finance Centre (Hong Kong, 2003) (Figure 6.14) and of cut corners in the World Trade Center Twin Towers (New York, 1972) (Figure 6.15).

6.1.2 Structure-based design

Buildings with symmetrical plan forms exhibit greater structural efficiency under lateral loads than buildings with asymmetrical plan forms (Ho, 2007). A building with a rectangular plan exhibits greater sensitivity to wind loads than buildings with circular, elliptical, or triangular plans (Ali and Armstrong, 1995). Thanks to the inherent strength of their geometry, these forms exhibit improved structural efficiency and help to reduce the construction cost. Examples include: the Marina City Towers (Chicago, 1964) (Figure 6.2), which has a cylindrical form; the Millennium Tower (Tokyo, 1993, proposed) (Figure 6.3), which has a conical form; and the U.S. Steel Tower (Pittsburgh, 1970) (Figure 6.16), which has a triangular plan.

FIGURE 6.16 U.S. Steel Tower, Pittsburgh, USA, 1970

Curved (crescent) or zigzag plans can also be used to increase the stiffness of the building against lateral loads. The behaviour of these forms against lateral loads is as effective as the behaviour exhibited by folded plates against vertical loads (Schueller, 1977). The City Hall Towers (Toronto, 1965) (Figure 6.17) and the Bow (Calgary, 2012), which have a curved plan form, are examples.

6.2 Structural design approach

The control of the dynamic response of a tall and flexible building can be achieved by increasing stiffness by the use of "shear-frame systems", "mega column systems", "mega core systems", "outriggered frame systems" and "tube systems" as a structural design approach which is discussed separately in Sections 3.6–3.10.

6.3 Mechanical design approach

In designing tall buildings, engineers assume a certain level of inherent damping in the structure when estimating the serviceability of the building under wind- and earthquake-induced lateral loads.

The inherent damping capacity of a building is affected by:

- the structural system
- the materials used in the structural system
- the cladding and non-structural elements such as interior and exterior partition walls
- the soil-structure interaction.

This is difficult to estimate and measure.

FIGURE 6.17 City Hall Towers, Toronto, Canada, 1965

Because the inherent damping capacity of buildings is often insufficient and difficult to estimate, auxiliary damping systems – the mechanical design approach – are used to reduce the effects of wind-induced vibration. These systems, known as "dampers", are mechanisms used to slow the lateral movement and reduce its magnitude and dampen the building sway by influencing the phases of force and displacement. Auxiliary damping systems can be divided into four groups:

- passive systems
- active systems
- semi-active systems
- hybrid systems.

Passive systems, which are systems that do not need an external power source, function by counteracting the building sway. These systems can be divided into two subgroups (Ali and Moon, 2007):

- energy-dissipating material-based damping systems
- auxiliary mass damping systems.

Energy-dissipating material-based damping systems are designed to dampen the dynamic motion of the building and are generally integrated with the structural system. Viscoelastic dampers (VEDs) are an example of this kind of system. The "elastic property" is the ability to return to an earlier state by storing energy, as with a coiled spring; the "viscous property" is the ability to flow like a thick liquid. The dampers, which combine elastic and viscous properties, dissipate the energy of the deformation caused by lateral forces by countering them with their viscoelastic behaviour and achieve damping by slowing down wind-induced vibration. Viscoelastic dampers were used in the World Trade Center Twin Towers (Figure 6.18).

Auxiliary mass systems are founded on the principle of creating drift in the opposite direction to the lateral drift of the building by creating a counteracting inertia force. Tuned mass dampers (TMDs) are an example of this kind of system. These dampers basically consist of mechanisms that control the function of a mass producing a counteracting inertia force and a mechanism to ensure the desired performance (Figure 6.19). The mass, which oscillates against wind-induced vibration, creates the counteracting inertia force. Generally it is located near the building top to obtain the best performance. In the Taipei 101, a 730–ton TMD was used near the top of the building (between the 87th and 92nd floors) (Figure 6.20). Tuned liquid dampers (TLDs) are also an example of auxiliary mass damping systems. These dampers consist of a liquid mass producing a counteracting inertia force and a mechanism to ensure the desired performance. The "sloshing" of the liquid mass creates an inertia force that counteracts the wind-induced vibration. Tuned liquid dampers have been considered for use in the Millennium Tower (proposed) (Figure 6.21).

Active systems, which aim to dampen wind-induced vibration, need an external energy source and are controlled by feedback from the structural responses. Active mass dampers (AMDs) are an example of this kind of system. These dampers resemble TMDs in their appearance, but while TMDs' ability to cope with a range of loads is

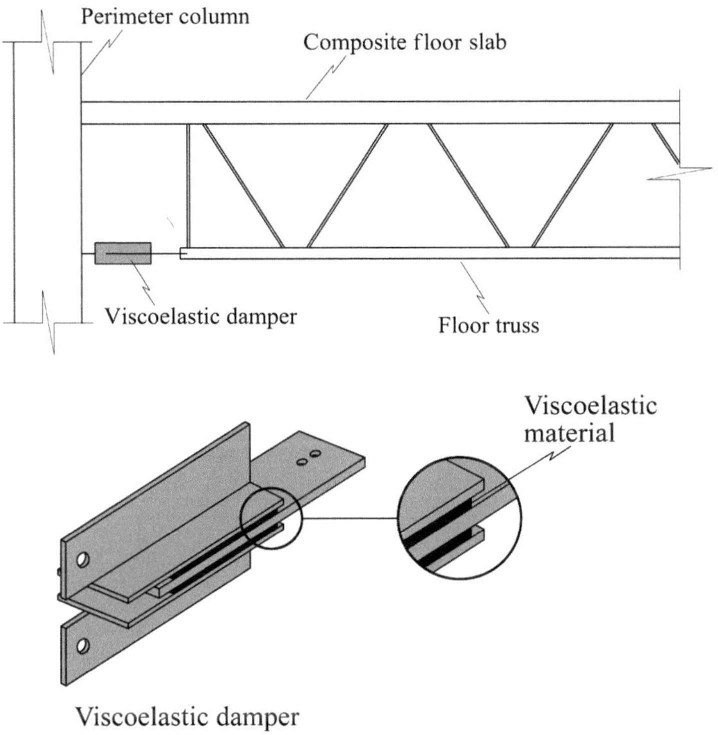

FIGURE 6.18 The use of viscoelastic dampers in the World Trade Center Twin Towers

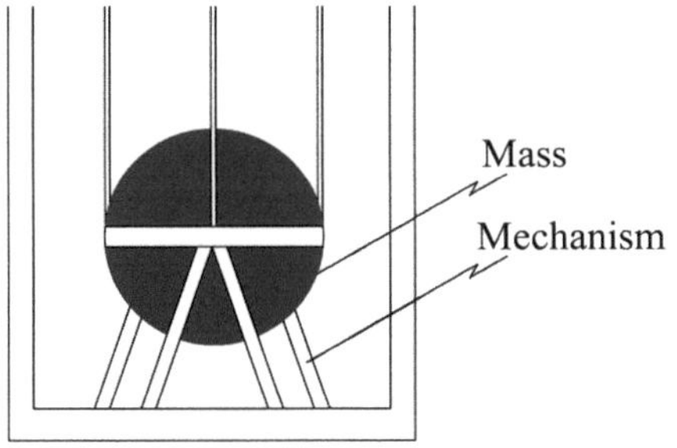

FIGURE 6.19 A tuned mass damper (TMD)

limited, AMDs can handle a much wider range of loads. Wind-induced vibration is monitored by a computer, which turns on the active mass dampers as necessary, damping the unwanted vibration. AMDs were used in the Applause Tower (Osaka, 1992) and the Nanjing TV Tower (Nanjing, 1993).

Although active systems are more efficient than passive systems, the possibility cannot be ignored that in extreme conditions they may be insufficient or impossible to activate because they require an external power source. For this reason, passive systems are preferable to active systems.

Semi-active systems are a subset of active systems, but need less external energy than active systems. Examples of these systems include semi-active impact dampers, adjustable tuned liquid dampers and controllable fluid dampers.

Hybrid systems are systems where active and passive systems, or semi-active and passive systems, are combined. Hybrid mass dampers (HMDs) are an example of this kind of system.

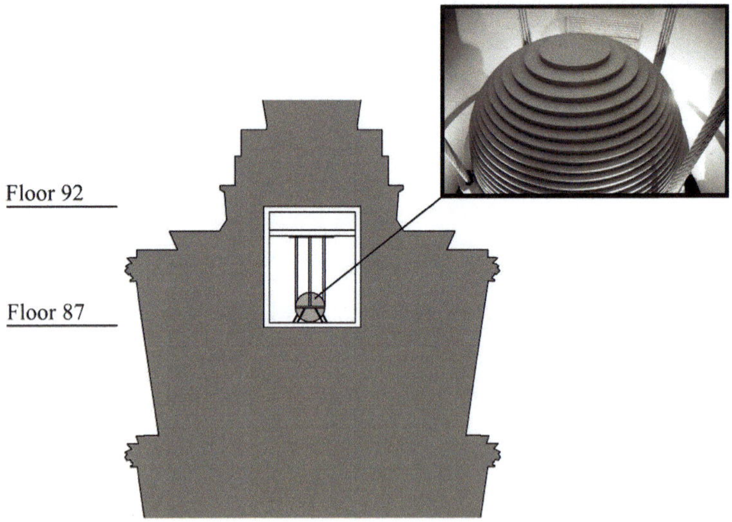

FIGURE 6.20 The use of a tuned mass damper in the Taipei 101

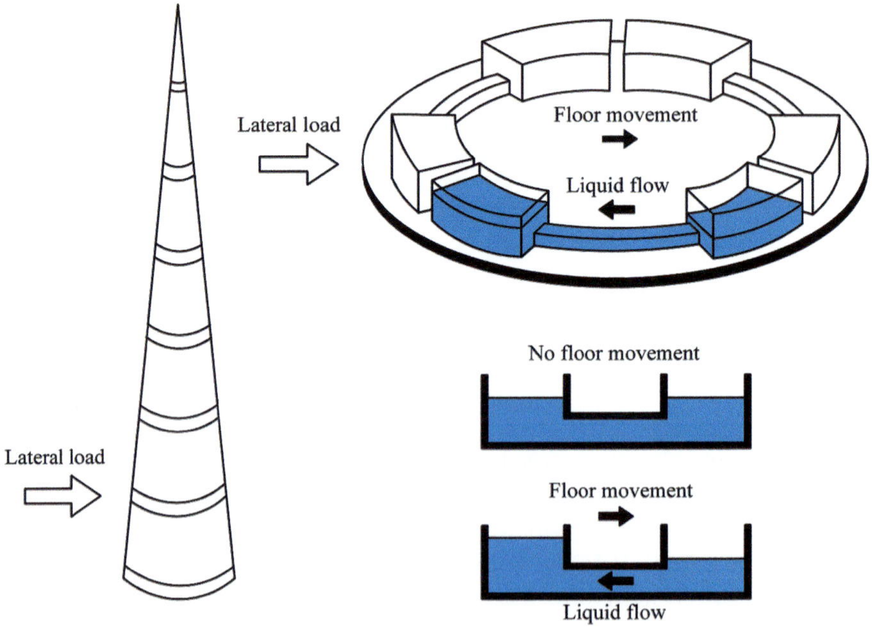

FIGURE 6.21 The use of a tuned liquid damper in the Millennium Tower (Kareem et al., 1999)

APPENDIX: EXAMPLES OF TALL BUILDINGS AND THEIR STRUCTURAL SYSTEMS

Shear-frame systems

OFFICIAL NAME	**Seagram Building**
LOCATION	New York, USA
BUILDING FUNCTION	Office
ARCHITECTURAL HEIGHT	157 m
NUMBER OF STOREYS	38
STATUS	Completed
COMPLETION	1958
ARCHITECT	Ludwig Mies van der Rohe; Kahn & Jacobs
STRUCTURAL ENGINEER	Severud Associates
STRUCTURAL SYSTEM	**Shear trussed frame + shear walled frame system/composite**

Shear trussed frame system

OFFICIAL NAME	**Empire State Building**
LOCATION	New York, USA
BUILDING FUNCTION	Office
ARCHITECTURAL HEIGHT	381 m
NUMBER OF STOREYS	102
STATUS	Completed
COMPLETION	1931
ARCHITECT	Shreve Lamb & Harmon Associates
STRUCTURAL ENGINEER	H.G. Balcom & Associates; Post and McCord; Strong & Jones Engineers
STRUCTURAL SYSTEM	**Shear trussed frame system/steel**

OFFICIAL NAME	**Chrysler Building**
LOCATION	New York, USA
BUILDING FUNCTION	Office
ARCHITECTURAL HEIGHT	319 m
NUMBER OF STOREYS	77
STATUS	Completed
COMPLETION	1930
ARCHITECT	William van Alen
STRUCTURAL ENGINEER	Ralph Squire & Sons
STRUCTURAL SYSTEM	**Shear trussed frame system/steel**

Shear walled frame system

OFFICIAL NAME	**Pirelli Building**
LOCATION	Milan, Italy
BUILDING FUNCTION	Office
ARCHITECTURAL HEIGHT	127 m
NUMBER OF STOREYS	32
STATUS	Completed
COMPLETION	1958
ARCHITECT	Gio Ponti
STRUCTURAL ENGINEER	Pier Luigi Nervi
STRUCTURAL SYSTEM	**Shear walled frame system/reinforced concrete**

OFFICIAL NAME	**CITIC Plaza**
LOCATION	Guangzhou, China
BUILDING FUNCTION	Office
ARCHITECTURAL HEIGHT	390 m
NUMBER OF STOREYS	80
STATUS	Completed
COMPLETION	1996
ARCHITECT	Dennis Lau & Ng Chun Man Architects & Engineers
STRUCTURAL ENGINEER	Maunsell AECOM Group
STRUCTURAL SYSTEM	**Shear walled frame system/reinforced concrete**

OFFICIAL NAME	**Q1**
LOCATION	Gold Coast, Australia
BUILDING FUNCTION	Residential
ARCHITECTURAL HEIGHT	323 m
NUMBER OF STOREYS	78
STATUS	Completed
COMPLETION	2005
ARCHITECT	Sunland Group
STRUCTURAL ENGINEER	Ove Arup & Partners
STRUCTURAL SYSTEM	**Shear walled frame system/reinforced concrete**

OFFICIAL NAME	**Kingdom Center**
LOCATION	Riyadh, Saudi Arabia
BUILDING FUNCTION	Mixed-use (Residential, hotel, office)
ARCHITECTURAL HEIGHT	302 m
NUMBER OF STOREYS	41
STATUS	Completed
COMPLETION	2002
ARCHITECT	Ellerbe Becket; Omrania & Associates
STRUCTURAL ENGINEER	Ove Arup & Partners
STRUCTURAL SYSTEM	**Shear walled frame system/reinforced concrete**

OFFICIAL NAME	**One Island East Centre**
LOCATION	Hong Kong, China
BUILDING FUNCTION	Office
ARCHITECTURAL HEIGHT	298 m
NUMBER OF STOREYS	69
STATUS	Completed
COMPLETION	2008
ARCHITECT	Wong & Ouyang
STRUCTURAL ENGINEER	Ove Arup & Partners
STRUCTURAL SYSTEM	**Shear walled frame system/reinforced concrete**

OFFICIAL NAME	**Comcast Center**
LOCATION	Philadelphia, USA
BUILDING FUNCTION	Office
ARCHITECTURAL HEIGHT	297 m
NUMBER OF STOREYS	57
STATUS	Completed
COMPLETION	2008
ARCHITECT	Robert A.M. Stern Architects; Kendall/ Heaton Associates
STRUCTURAL ENGINEER	Thornton Tomasetti
STRUCTURAL SYSTEM	**Shear walled frame system/ reinforced concrete**

OFFICIAL NAME	**311 South Wacker Drive**
LOCATION	Chicago, USA
BUILDING FUNCTION	Office
ARCHITECTURAL HEIGHT	293 m
NUMBER OF STOREYS	65
STATUS	Completed
COMPLETION	1990
ARCHITECT	Kohn Pedersen Fox Associates; HKS Architects
STRUCTURAL ENGINEER	Brockette Davis Drake
STRUCTURAL SYSTEM	**Shear walled frame system/ reinforced concrete**

OFFICIAL NAME	**Al Faisaliah Center**
LOCATION	Riyadh, Saudi Arabia
BUILDING FUNCTION	Office
ARCHITECTURAL HEIGHT	267 m
NUMBER OF STOREYS	30
STATUS	Completed
COMPLETION	2000
ARCHITECT	Foster + Partners
STRUCTURAL ENGINEER	Buro Happold; Sendai Eversendai Engineering Group
STRUCTURAL SYSTEM	**Shear walled frame system/ reinforced concrete**

OFFICIAL NAME	**Highcliff**
LOCATION	Hong Kong, China
BUILDING FUNCTION	Residential
ARCHITECTURAL HEIGHT	252 m
NUMBER OF STOREYS	73
STATUS	Completed
COMPLETION	2003
ARCHITECT	Dennis Lau & Ng Chun Man Architects & Engineers
STRUCTURAL ENGINEER	Magnusson Klemencic Associates; Canwest Consultants Limited
STRUCTURAL SYSTEM	**Shear walled frame system/ reinforced concrete**

OFFICIAL NAME	**Bahrain World Trade Center**
LOCATION	Manama, Kingdom of Bahrain
BUILDING FUNCTION	Mixed-use (Office, commercial)
ARCHITECTURAL HEIGHT	240 m
NUMBER OF STOREYS	45
STATUS	Completed
COMPLETION	2008
ARCHITECT	WS Atkins & Partners
STRUCTURAL ENGINEER	WS Atkins & Partners
STRUCTURAL SYSTEM	**Shear walled frame system/ reinforced concrete**

OFFICIAL NAME	**Lake Point Tower**
LOCATION	Chicago, USA
BUILDING FUNCTION	Residential
ARCHITECTURAL HEIGHT	197 m
NUMBER OF STOREYS	70
STATUS	Completed
COMPLETION	1968
ARCHITECT	Schipporeit-Heinrich Associates; Graham, Anderson, Probst & White
STRUCTURAL ENGINEER	William Schmidt & Associates
STRUCTURAL SYSTEM	**Shear walled frame system/ reinforced concrete**

OFFICIAL NAME	**Strata**
LOCATION	London, UK
BUILDING FUNCTION	Residential
ARCHITECTURAL HEIGHT	148 m
NUMBER OF STOREYS	43
STATUS	Completed
COMPLETION	2010
ARCHITECT	BFLS
STRUCTURAL ENGINEER	WSP Group
STRUCTURAL SYSTEM	**Shear walled frame system/ reinforced concrete**

OFFICIAL NAME	**Cook County Administration Building (formerly Brunswick Building)**
LOCATION	Chicago, USA
BUILDING FUNCTION	Office
ARCHITECTURAL HEIGHT	145 m
NUMBER OF STOREYS	35
STATUS	Completed
COMPLETION	1964
ARCHITECT	Bruce Graham (SOM)
STRUCTURAL ENGINEER	Fazlur Rahman Khan (SOM)
STRUCTURAL SYSTEM	**Shear walled frame system/ reinforced concrete**

Mega column (mega frame, space truss) systems

OFFICIAL NAME	**The Center**
LOCATION	Hong Kong, China
BUILDING FUNCTION	Office
ARCHITECTURAL HEIGHT	346 m
NUMBER OF STOREYS	73
STATUS	Completed
COMPLETION	1998
ARCHITECT	Dennis Lau & Ng Chun Man Architects & Engineers
STRUCTURAL ENGINEER	Maunsell AECOM Group
STRUCTURAL SYSTEM	**Mega column system/composite**

OFFICIAL NAME	**Commerzbank Tower**
LOCATION	Frankfurt, Germany
BUILDING FUNCTION	Office
ARCHITECTURAL HEIGHT	259 m
NUMBER OF STOREYS	56
STATUS	Completed
COMPLETION	1997
ARCHITECT	Foster + Partners
STRUCTURAL ENGINEER	Ove Arup & Partners; Krebs und Kiefer
STRUCTURAL SYSTEM	**Mega column system/composite**

Mega core systems

OFFICIAL NAME	**Aspire Tower**
LOCATION	Doha, Qatar
BUILDING FUNCTION	Mixed-use (Hotel, office)
ARCHITECTURAL HEIGHT	300 m
NUMBER OF STOREYS	36
STATUS	Completed
COMPLETION	2007
ARCHITECT	Hadi Simaan
STRUCTURAL ENGINEER	Ove Arup & Partners
STRUCTURAL SYSTEM	**Mega core system/reinforced concrete**

OFFICIAL NAME	**8 Shenton Way**
LOCATION	Singapore, Singapore
BUILDING FUNCTION	Office
ARCHITECTURAL HEIGHT	235 m
NUMBER OF STOREYS	52
STATUS	Completed
COMPLETION	1986
ARCHITECT	The Stubbins Associates; Architects 61 Private
STRUCTURAL ENGINEER	Ove Arup & Partners
STRUCTURAL SYSTEM	**Mega core system/reinforced concrete**

OFFICIAL NAME	**HSB Turning Torso**
LOCATION	Malmö, Sweden
BUILDING FUNCTION	Residential
ARCHITECTURAL HEIGHT	190 m
NUMBER OF STOREYS	57
STATUS	Completed
COMPLETION	2005
ARCHITECT	Santiago Calatrava
STRUCTURAL ENGINEER	SWECO
STRUCTURAL SYSTEM	**Mega core system/reinforced concrete**

Outriggered frame systems

OFFICIAL NAME	**Burj Khalifa (formerly Burj Dubai)**
LOCATION	Dubai, U.A.E
BUILDING FUNCTION	Mixed-use (Hotel, residential, office)
ARCHITECTURAL HEIGHT	828 m
NUMBER OF STOREYS	163
STATUS	Completed
COMPLETION	2010
ARCHITECT	Skidmore, Owings & Merrill (SOM); Hyder Consulting
STRUCTURAL ENGINEER	William F. Baker (SOM)
STRUCTURAL SYSTEM	**Outriggered frame system/ reinforced concrete**

OFFICIAL NAME	**Shanghai Tower**
LOCATION	Shanghai, China
BUILDING FUNCTION	Mixed-use (Hotel, office)
ARCHITECTURAL HEIGHT	632 m
NUMBER OF STOREYS	121
STATUS	Under construction
COMPLETION	2014 (estimated)
ARCHITECT	Gensler
STRUCTURAL ENGINEER	Thornton Tomasetti
STRUCTURAL SYSTEM	**Outriggered frame system/composite**

OFFICIAL NAME	**Taipei 101**
LOCATION	Taipei, Taiwan
BUILDING FUNCTION	Office
ARCHITECTURAL HEIGHT	508 m
NUMBER OF STOREYS	101
STATUS	Completed
COMPLETION	2004
ARCHITECT	C.Y. Lee & Partners
STRUCTURAL ENGINEER	Thornton Tomasetti; Evergreen Engineering
STRUCTURAL SYSTEM	**Outriggered frame system/composite**

OFFICIAL NAME	**Shanghai World Financial Center**
LOCATION	Shanghai, China
BUILDING FUNCTION	Mixed-use (Office, hotel)
ARCHITECTURAL HEIGHT	492 m
NUMBER OF STOREYS	101
STATUS	Completed
COMPLETION	2008
ARCHITECT	Kohn Pedersen Fox Associates; Irie Miyake Architects and Engineers
STRUCTURAL ENGINEER	Leslie E. Robertson Associates
STRUCTURAL SYSTEM	**Outriggered frame system/composite**

OFFICIAL NAME	**International Commerce Centre (ICC)**
LOCATION	Hong Kong, China
BUILDING FUNCTION	Mixed-use (Hotel, office)
ARCHITECTURAL HEIGHT	484 m
NUMBER OF STOREYS	108
STATUS	Completed
COMPLETION	2010
ARCHITECT	Kohn Pedersen Fox Associates
STRUCTURAL ENGINEER	Ove Arup & Partners
STRUCTURAL SYSTEM	**Outriggered frame system/composite**

OFFICIAL NAME	**The Petronas Twin Towers**
LOCATION	Kuala Lumpur, Malaysia
BUILDING FUNCTION	Office
ARCHITECTURAL HEIGHT	452 m
NUMBER OF STOREYS	88
STATUS	Completed
COMPLETION	1998
ARCHITECT	César Pelli & Associates
STRUCTURAL ENGINEER	Thornton Tomasetti; Ranhill Bersekutu
STRUCTURAL SYSTEM	**Outriggered frame system/ reinforced concrete**

OFFICIAL NAME	**Zifeng Tower (formerly Nanjing Greenland Financial Center)**
LOCATION	Nanjing, China
BUILDING FUNCTION	Mixed-use (Hotel, office)
ARCHITECTURAL HEIGHT	450 m
NUMBER OF STOREYS	66
STATUS	Completed
COMPLETION	2010
ARCHITECT	Skidmore, Owings & Merrill (SOM)
STRUCTURAL ENGINEER	Skidmore, Owings & Merrill (SOM)
STRUCTURAL SYSTEM	**Outriggered frame system/composite**

OFFICIAL NAME	**Trump International Hotel & Tower**
LOCATION	Chicago, USA
BUILDING FUNCTION	Mixed-use (Residential, hotel)
ARCHITECTURAL HEIGHT	423 m
NUMBER OF STOREYS	98
STATUS	Completed
COMPLETION	2009
ARCHITECT	Skidmore, Owings & Merrill (SOM)
STRUCTURAL ENGINEER	Skidmore, Owings & Merrill (SOM)
STRUCTURAL SYSTEM	**Outriggered frame system/ reinforced concrete**

OFFICIAL NAME	**Jin Mao Building**
LOCATION	Shanghai, China
BUILDING FUNCTION	Mixed-use (Office, hotel)
ARCHITECTURAL HEIGHT	421 m
NUMBER OF STOREYS	88
STATUS	Completed
COMPLETION	1999
ARCHITECT	Skidmore, Owings & Merrill (SOM)
STRUCTURAL ENGINEER	Skidmore, Owings & Merrill (SOM)
STRUCTURAL SYSTEM	**Outriggered frame system/composite**

OFFICIAL NAME	**Two International Finance Centre**
LOCATION	Hong Kong, China
BUILDING FUNCTION	Office
ARCHITECTURAL HEIGHT	412 m
NUMBER OF STOREYS	88
STATUS	Completed
COMPLETION	2003
ARCHITECT	César Pelli & Associates; Rocco Design
STRUCTURAL ENGINEER	Ove Arup & Partners
STRUCTURAL SYSTEM	**Outriggered frame system/composite**

OFFICIAL NAME	**Shun Hing Square**
LOCATION	Shenzhen, China
BUILDING FUNCTION	Office
ARCHITECTURAL HEIGHT	384 m
NUMBER OF STOREYS	69
STATUS	Completed
COMPLETION	1996
ARCHITECT	K.Y. Cheung Design Associates; American Design Associates
STRUCTURAL ENGINEER	Maunsell AECOM Group; Nippon Steel Corp; Leslie E. Robertson Associates
STRUCTURAL SYSTEM	**Outriggered frame system/composite**

OFFICIAL NAME	**New York Times Tower**
LOCATION	New York, USA
BUILDING FUNCTION	Office
ARCHITECTURAL HEIGHT	319 m
NUMBER OF STOREYS	52
STATUS	Completed
COMPLETION	2007
ARCHITECT	FXFOWLE; Renzo Piano Building Workshop
STRUCTURAL ENGINEER	Thornton Tomasetti
STRUCTURAL SYSTEM	**Outriggered frame system/steel**

OFFICIAL NAME	**Menara Telekom**
LOCATION	Kuala Lumpur, Malaysia
BUILDING FUNCTION	Office
ARCHITECTURAL HEIGHT	310 m
NUMBER OF STOREYS	55
STATUS	Completed
COMPLETION	2001
ARCHITECT	Hijjas Kasturi Associates
STRUCTURAL ENGINEER	Ranhill Bersekutu
STRUCTURAL SYSTEM	**Outriggered frame system/ reinforced concrete**

OFFICIAL NAME	**Eureka Tower**
LOCATION	Melbourne, Australia
BUILDING FUNCTION	Residential
ARCHITECTURAL HEIGHT	297 m
NUMBER OF STOREYS	91
STATUS	Completed
COMPLETION	2006
ARCHITECT	Fender Katsalidis
STRUCTURAL ENGINEER	Connell Mott MacDonald
STRUCTURAL SYSTEM	**Outriggered frame system/ reinforced concrete**

OFFICIAL NAME	**Plaza 66**
LOCATION	Shanghai, China
BUILDING FUNCTION	Office
ARCHITECTURAL HEIGHT	288 m
NUMBER OF STOREYS	66
STATUS	Completed
COMPLETION	2001
ARCHITECT	Kohn Pedersen Fox Associates; Frank CY Feng Architects & Associates
STRUCTURAL ENGINEER	Thornton Tomasetti
STRUCTURAL SYSTEM	**Outriggered frame system/ reinforced concrete**

OFFICIAL NAME	**Cheung Kong Centre**
LOCATION	Hong Kong, China
BUILDING FUNCTION	Office
ARCHITECTURAL HEIGHT	283 m
NUMBER OF STOREYS	63
STATUS	Completed
COMPLETION	1999
ARCHITECT	César Pelli & Associates; Leo A. Daly; Hsin Yieh Architects & Associates
STRUCTURAL ENGINEER	Ove Arup & Partners
STRUCTURAL SYSTEM	**Outriggered frame system/composite**

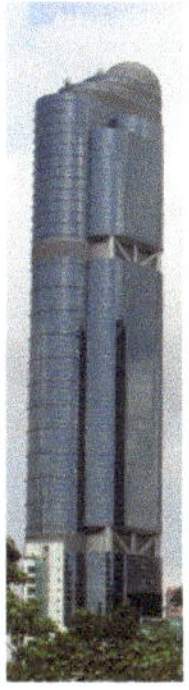

OFFICIAL NAME	**Langham Place Office Tower**
LOCATION	Hong Kong, China
BUILDING FUNCTION	Office
ARCHITECTURAL HEIGHT	255 m
NUMBER OF STOREYS	59
STATUS	Completed
COMPLETION	2004
ARCHITECT	Wong & Ouyang
STRUCTURAL ENGINEER	Ove Arup & Partners
STRUCTURAL SYSTEM	**Outriggered frame system/ reinforced concrete**

OFFICIAL NAME	**World Tower**
LOCATION	Sydney, Australia
BUILDING FUNCTION	Residential
ARCHITECTURAL HEIGHT	230 m
NUMBER OF STOREYS	73
STATUS	Completed
COMPLETION	2004
ARCHITECT	Nation Fender Katsalidis
STRUCTURAL ENGINEER	Connell Mott MacDonald
STRUCTURAL SYSTEM	**Outriggered frame system/ reinforced concrete**

Tube systems

Framed-tube systems

OFFICIAL NAME	**World Trade Center Twin Towers (WTC I – II)**
LOCATION	New York, USA
BUILDING FUNCTION	Office
ARCHITECTURAL HEIGHT	415/417 m
NUMBER OF STOREYS	110
STATUS	Demolished after collapse following fire and airplane impact (2001)
COMPLETION	1972 (WTC I), 1973 (WTC II)
ARCHITECT	Minoru Yamazaki & Associates; Emery Roth & Sons
STRUCTURAL ENGINEER	Leslie E. Robertson Associates
STRUCTURAL SYSTEM	**Framed-tube system/steel**

OFFICIAL NAME	**Water Tower Place**
LOCATION	Chicago, USA
BUILDING FUNCTION	Mixed-use (Residential, hotel, retail)
ARCHITECTURAL HEIGHT	262 m
NUMBER OF STOREYS	74
STATUS	Completed
COMPLETION	1976
ARCHITECT	Loebl Schlossman Dart & Hackl; C.F. Murphy Associates
STRUCTURAL ENGINEER	C.F. Murphy Associates
STRUCTURAL SYSTEM	**Framed-tube system/ reinforced concrete**

OFFICIAL NAME	**Olympia Centre**
LOCATION	Chicago, USA
BUILDING FUNCTION	Mixed-use (Residential, office)
ARCHITECTURAL HEIGHT	223 m
NUMBER OF STOREYS	63
STATUS	Completed
COMPLETION	1986
ARCHITECT	Skidmore, Owings & Merrill (SOM)
STRUCTURAL ENGINEER	Skidmore, Owings & Merrill (SOM)
STRUCTURAL SYSTEM	**Framed-tube system/ reinforced concrete**

OFFICIAL NAME	**First Canadian Centre**
LOCATION	Calgary, Canada
BUILDING FUNCTION	Office
ARCHITECTURAL HEIGHT	167 m
NUMBER OF STOREYS	41
STATUS	Completed
COMPLETION	1982
ARCHITECT	Bregman + Hamann Architects
STRUCTURAL ENGINEER	Read Jones Christoffersen Ltd.
STRUCTURAL SYSTEM	**Framed-tube system/ reinforced concrete**

OFFICIAL NAME	**Torre Agbar**
LOCATION	Barcelona, Spain
BUILDING FUNCTION	Office
ARCHITECTURAL HEIGHT	144 m
NUMBER OF STOREYS	33
STATUS	Completed
COMPLETION	2004
ARCHITECT	Ateliers Jean Nouvel
STRUCTURAL ENGINEER	Robert Brufau y Asociados; Obiol, Moya i Associats
STRUCTURAL SYSTEM	**Framed-tube system/ reinforced concrete**

OFFICIAL NAME	**The Plaza on Dewitt (Dewitt-Chestnut Apartments)**
LOCATION	Chicago, USA
BUILDING FUNCTION	Residential
ARCHITECTURAL HEIGHT	120 m
NUMBER OF STOREYS	43
STATUS	Completed
COMPLETION	1966
ARCHITECT	Myron Goldsmith (SOM)
STRUCTURAL ENGINEER	Fazlur Rahman Khan (SOM)
STRUCTURAL SYSTEM	**Framed-tube system/ reinforced concrete**

OFFICIAL NAME	**Guangzhou International Finance Center**
LOCATION	Guangzhou, China
BUILDING FUNCTION	Mixed-use (Hotel, office)
ARCHITECTURAL HEIGHT	439 m
NUMBER OF STOREYS	103
STATUS	Completed
COMPLETION	2010
ARCHITECT	Wilkinson Eyre
STRUCTURAL ENGINEER	Ove Arup & Partners; Architecture Design Institute of South China University of Technology
STRUCTURAL SYSTEM	**Trussed-tube system (diagrid-framed-tube system)/steel**

OFFICIAL NAME	**Hearst Magazine Tower**
LOCATION	New York, USA
BUILDING FUNCTION	Office
ARCHITECTURAL HEIGHT	182 m
NUMBER OF STOREYS	46
STATUS	Completed
COMPLETION	2006
ARCHITECT	Foster + Partners
STRUCTURAL ENGINEER	WSP Cantor Seinuk
STRUCTURAL SYSTEM	**Trussed-tube system (diagrid-framed-tube system)/composite**

OFFICIAL NAME	**30 St Mary Axe**
LOCATION	London, United Kingdom
BUILDING FUNCTION	Office
ARCHITECTURAL HEIGHT	180 m
NUMBER OF STOREYS	41
STATUS	Completed
COMPLETION	2004
ARCHITECT	Foster + Partners
STRUCTURAL ENGINEER	Ove Arup & Partners
STRUCTURAL SYSTEM	**Trussed-tube system (diagrid-framed-tube system)/steel**

OFFICIAL NAME	**COR Building**
LOCATION	Miami, USA
BUILDING FUNCTION	Mixed-use (Residential, office)
ARCHITECTURAL HEIGHT	118 m
NUMBER OF STOREYS	40
STATUS	Project pending
COMPLETION	
ARCHITECT	Oppenheim Architecture+Design
STRUCTURAL ENGINEER	Ysrael A. Seinuk, PC
STRUCTURAL SYSTEM	**Trussed-tube system (diagrid-framed-tube system)/ reinforced concrete**

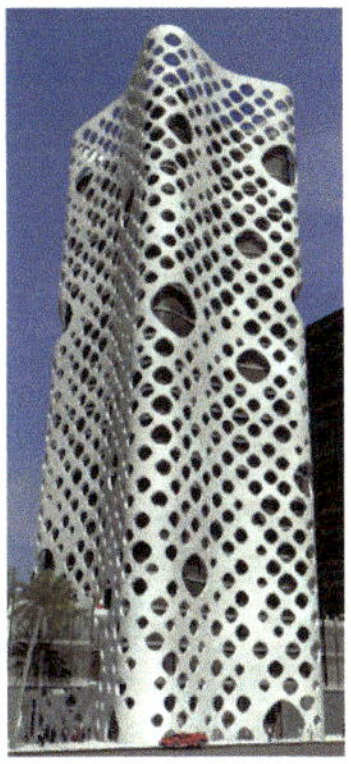

OFFICIAL NAME	**O-14**
LOCATION	Dubai, U.A.E
BUILDING FUNCTION	Office
ARCHITECTURAL HEIGHT	106 m
NUMBER OF STOREYS	22
STATUS	Completed
COMPLETION	2010
ARCHITECT	Reiser + Umemoto RUR Architecture P.C.
STRUCTURAL ENGINEER	Ysrael A. Seinuk, PC
STRUCTURAL SYSTEM	**Trussed-tube system (diagrid-framed-tube system)/ reinforced concrete**

Trussed-tube systems

OFFICIAL NAME	**Bank of China Tower**
LOCATION	Hong Kong, China
BUILDING FUNCTION	Office
ARCHITECTURAL HEIGHT	367 m
NUMBER OF STOREYS	72
STATUS	Completed
COMPLETION	1990
ARCHITECT	I.M. Pei & Partners; Sherman Kung & Associates Architects
STRUCTURAL ENGINEER	Leslie E. Robertson Associates; Valentine, Laurie, and Davis
STRUCTURAL SYSTEM	**Trussed-tube system/mega frame system/space truss system/composite**

OFFICIAL NAME	**John Hancock Center**
LOCATION	Chicago, USA
BUILDING FUNCTION	Mixed-use (Residential, office)
ARCHITECTURAL HEIGHT	344 m
NUMBER OF STOREYS	100
STATUS	Completed
COMPLETION	1969
ARCHITECT	Bruce Graham (SOM)
STRUCTURAL ENGINEER	Fazlur Rahman Khan (SOM)
STRUCTURAL SYSTEM	**Trussed-tube system/steel**

OFFICIAL NAME	**Citigroup Center**
LOCATION	New York, USA
BUILDING FUNCTION	Office
ARCHITECTURAL HEIGHT	279 m
NUMBER OF STOREYS	59
STATUS	Completed
COMPLETION	1977
ARCHITECT	Stubbins Associates; Emery Roth & Sons
STRUCTURAL ENGINEER	LeMessurier Consultants; The Office of James Ruderman
STRUCTURAL SYSTEM	**Trussed-tube system/composite**

OFFICIAL NAME	**CCTV Headquarters**
LOCATION	Beijing, China
BUILDING FUNCTION	Office
ARCHITECTURAL HEIGHT	234 m
NUMBER OF STOREYS	49
STATUS	Completed
COMPLETION	2011
ARCHITECT	Office for Metropolitan Architecture
STRUCTURAL ENGINEER	East China Architectural Design and Research; Ove Arup & Partners
STRUCTURAL SYSTEM	**Trussed-tube system/composite**

OFFICIAL NAME	**Onterie Center**
LOCATION	Chicago, USA
BUILDING FUNCTION	Mixed-use (Residential, office)
ARCHITECTURAL HEIGHT	174 m
NUMBER OF STOREYS	58
STATUS	Completed
COMPLETION	1986
ARCHITECT	Skidmore, Owings & Merrill (SOM)
STRUCTURAL ENGINEER	Fazlur Rahman Khan (SOM)
STRUCTURAL SYSTEM	**Trussed-tube system/ reinforced concrete**

OFFICIAL NAME	**780 Third Avenue Building**
LOCATION	New York, USA
BUILDING FUNCTION	Office
ARCHITECTURAL HEIGHT	174 m
NUMBER OF STOREYS	50
STATUS	Completed
COMPLETION	1983
ARCHITECT	Skidmore, Owings & Merrill (SOM)
STRUCTURAL ENGINEER	Rosenwasser/Grossman Consulting Engineers, P.C.
STRUCTURAL SYSTEM	**Trussed-tube system/ reinforced concrete**

Bundled-tube systems

OFFICIAL NAME	**Willis Tower (formerly Sears Tower)**
LOCATION	Chicago, USA
BUILDING FUNCTION	Office
ARCHITECTURAL HEIGHT	442 m
NUMBER OF STOREYS	108
STATUS	Completed
COMPLETION	1974
ARCHITECT	Bruce Graham (SOM)
STRUCTURAL ENGINEER	Fazlur Rahman Khan (SOM)
STRUCTURAL SYSTEM	**Bundled-tube system/steel**

OFFICIAL NAME	**Wachovia Financial Center**
LOCATION	Miami, USA
BUILDING FUNCTION	Office
ARCHITECTURAL HEIGHT	233 m
NUMBER OF STOREYS	55
STATUS	Completed
COMPLETION	1983
ARCHITECT	Skidmore, Owings & Merrill (SOM)
STRUCTURAL ENGINEER	Skidmore, Owings & Merrill (SOM)
STRUCTURAL SYSTEM	**Bundled-tube system/composite**

OFFICIAL NAME	**One Magnificent Mile**
LOCATION	Chicago, USA
BUILDING FUNCTION	Mixed-use (Residential, office)
ARCHITECTURAL HEIGHT	205 m
NUMBER OF STOREYS	57
STATUS	Completed
COMPLETION	1983
ARCHITECT	Skidmore, Owings & Merrill (SOM)
STRUCTURAL ENGINEER	Fazlur Rahman Khan (SOM)
STRUCTURAL SYSTEM	**Bundled-tube system/ reinforced concrete**

BIBLIOGRAPHY

Abdelrazaq, A., Validating the Dynamics of the Burj Khalifa, *CTBUH Journal*, CTBUH Technical Paper, Issue II, 2011.

Abdelrazaq, A., Kim, K.J. and Kim, J.H., *Brief on the Construction Planning of The Burj Dubai Project*, UAE, CTBUH 8th World Congress 2008, Dubai, 2008.

ACI SP-97, *Analysis and Design of High-Rise Concrete Buildings*, U.S.A, 1989.

Agnoletto, M., *Renzo Piano*, Milan: Motta Architettura srl, 2006.

Ali, M. and Armstrong, P., *Architecture of Tall Buildings*, Council on Tall Buildings and Urban Habitat Committee (CTBUH), New York: McGraw-Hill Book Company, 1995.

Ali, M. and Moon, K., Structural Developments in Tall Buildings: Current Trends and Future Prospects, Invited Review Paper, *Architectural Science Review*, Vol. 50, No. 3, pp. 205–223, 2007.

Bailey, P., Bridges, H., Cross, P., Del Mese, G., Smith C., Walsh, S. and Wise, C., Commerzbank, Frankfurt, *The Arup Journal*, No 2, 1997.

Baker, W., *The World's Tallest Building, Burj Dubai, UAE*, Proceedings of the CTBUH 2004 Seoul Conference, pp. 1168–1169, Seoul, 2004.

Baker, W.F., Korista, D.S. and Novak, L.C., Burj Dubai: Engineering the World's Tallest Building, *The Structural Design of Tall and Special Buildings*, Vol. 16, pp. 361–375, 2007.

Baker, W.F., Korista, D.S. and Novak, L.C., *Engineering the World's Tallest – Burj Dubai*, Proceedings of the CTBUH 8th World Congress 2008, Dubai, 2008.

Banham, R., *Age of the Masters*, Architectural Press Ltd., pp. 116–117, 1977.

Beedle, L.S., *What's a Tall Building?*, ASCE Annual and Environmental Engineering Meeting, St. Louis, MO, October, ASCE, Preprint no. 1553 (M20), New York, 1971.

Bellini, O.E. and Daglio, L., *Bahrain World Trade Center*, Building Arabia, White Star Publishers, pp. 36–40, Vercelli, 2009.

Bennett, D., *Skyscrapers: Form & Function*, Simon & Schuster Ltd., New York, 1995.

Binder, G., *101 of the World's Tallest Buildings*, pp. 204–205, Victoria: Images Publishing Group, 2006.

Boake, T. M., Green Super Buildings, *Alternatives Journal*, November/December, 30/5, 2004.

Buchanan, P., Reinventing the Sky-Scrape, *A and U*. No 329, February, 30–67, 1998.

Buchanan, P., *Renzo Piano Building Workshop*, London: Phaidon Press Limited, 2008.

BWTC Scores World First, *Gulf Construction Magazine*, pp. 71–76, May 2008.

Callow, J.A., Krall, E. and Scarangello, T.Z., Inside Out, *Modern Steel Construction*, January 2009.

Carroll, C., Cross, P., Duan, X., Gibbons, C., Ho, G., Kwok, M., Lawson, R., Lee, A., Luong, A., McGowan, R. and Pope, C., CCTV Headquarters, Beijing, China: Structural Engineering Design and Approvals, *The Arup Journal*, No. 2, 2005.

Carroll, C., Cross, P., Duan, X., Gibbons, C., Ho, G., Kwok, M., Lawson, R., Lee, A., Li, R., Luong, A., McGowan, R. and Pope, C., Case Study: CCTV Building, *CTBUH Journal*, Issue: 3, 2008.

Carroll, C., Gibbons, C., Ho, G., Kwok, M., Lawson, R., Lee, A., Li, R., Luong, A., McGowan, R. and Pope, C., CCTV Headquarters, Beijing, China: Building the Structure, *The Arup Journal*, No. 2, 2008.

Cement and Concrete Association of Australia, Mix Magazine, *High Strength Concrete: Recent Developments with 100 MPa at The Eureka Tower,* No. 12, pp. 2–5, July 2003.

Chen, A., Shanghai World Financial Center, *Architectural Record* 05.09, pp. 184–190, 2009.

Chen, L.C., Chen, C.C., Chen, C.Y., Yeh, S.W., Zen, S.K. and Chou, J.H., The Wind Features of Taipei 101 Financial Building, *Journal of Wind Engineering,* No. 108, pp. 725–728, 2006.

Chikaher, G. and Hirst, J., Aspire Tower, Doha, Qatar, *The Arup Journal* 2/2007, pp. 3–13, 2007.

Chiorino, M. and Chiorino, C., *Pier Luigi Nervi: Architecture as Challenge*, Milan, 2010. (Editors: Olmo Carlo, Cristiana Chiorino, Exhibition held at Palazzo Giustinian Lolin, Venice, p. 237.)

Colaco, J.P., *Structural Systems for Tall Apartment Towers*, CTBUH 7th World Congress 2005, New York, 2005.

Corley, W.G., *Structural Investigation of the New York World Trade Center Collapse*, Proceedings of the CIB-CTBUH International Conference on Tall Buildings, Malaysia, 2003.

CTBUH, Council on Tall Buildings and Urban Habitat, Illinois Institute of Technology, S.R. Crown Hall, 3360 South State Street, Chicago, Illinois, USA, www.ctbuh.org.

Curtis, W. J. R., *Modern Architecture Since 1900*, Phaidon Press Ltd., London, p. 478, 1996.

Davenport, A.G., The Response of Super Tall Buildings to Wind, *Second Century of the Skyscraper, Council on Tall Buildings and Urban Habitat*, pp. 705–725, 1988.

Dean, B., Martin, O., Emery, D. and Chancellor, P., *Tall Building Design Innovations in Australia*, pp. 393–399, CTBUH 6th World Congress 2001, Spon Press, Taylor & Francis Group, London and New York, 2001.

Domus, *La Torre Pirelli a Milano*, No. 348, pp. 1–3, 1958.

Dutton, R. and Isyumov, N., Reduction of Tall Building Motion by Aerodynamic Treatments, *Journal of Wind Engineering and Industrial Aerodynamics*, Vol. 36, 1990.

Emporis, *Emporis Corporation, A Global Building Information Company*, Theodor-Heuss-Allee 2, 60486 Frankfurt, Germany, www.emporis.com.

Fazlur Rahman Khan – John Hancock Center, http://khan.princeton.edu, accessed May 2012.

Fazlur Rahman Khan – Sears Tower, http://khan.princeton.edu, accessed May 2012.

Federal Emergency Management Agency (2001) Federal Insurance and Mitigation Administration, *World Trade Center Building Performance Study*, FEMA Report, 2001.

Forest City Enterprises, Inc., *Our Newest Value Creation Story: The New York Times Building, 2007 Summary Annual Report and Supplemental Package,* 2007.

Foster+Partners, *Commerzbank Headquarters Frankfurt, Germany 1991–1997*, http://www.fosterandpartners.com/projects/commerzbank-headquarters, accessed June 2013.

Freyssinet Magazine, *Two Prestressed Skyscrapers,* No. 213, p. 16, January/April 2002.

From Twisting to Turning Torso, *Kinnarps Magazine*, No. 5, 2006.

Fukuda, R., Fukuzawa, E., Koshika, N., Morikawa, H., Yoshikai, S., Makibe, K. and Mihara, Y., *Analyses of Structural Responses and Damages of World Trade Center Tower 2 Due to Impact of Aircraft Crashed*, Proceedings of the CIB-CTBUH International Conference on Tall Buildings, Malaysia, 2003.

Giblin, J.C., *The Skyscraper Book*, Thomas Y Crowell Junior Books, New York, 1981.

Godfrey, B., *Multi-Storey Buildings in Steel,* Second Edition, Collins, London, England, pp. 162–164, 1985.

Green, D., Lau, B., Lyall, I., Pritchard, D., Pullen, J. and Ryan, C., CCTV Headquarters, Beijing, China: Services Engineering Design, *The Arup Journal*, No. 3, 2005.

Gu, M. and Quan, Y., Across-wind Loads of Typical Tall Buildings, *Journal of Wind Engineering and Industrial Aerodynamics*, Vol. 92, pp. 1147–1165, 2004.

Hadenius, P., The World's Tallest Building for Now, *Technology Review*, Massachusetts Institute of Technology/MIT, August 2004.

Harbert, L., Home Insurance Building – The First Skyscraper, *Journal of American Society of Civil Engineers (ASCE)*, Vol. 43, No. 2, pp. 1–2, 2002.

Harris, B., *The World Trade Center: A Tribute*, Running Press Book Publishers, The United States, 2001.

Hedrick, M. S., Horst, K., Lehman, C. and Perez, A., *The New York Times Building*, The Pennsylvania State University, Pennsylvania, 2010.

Ho, P.H.K, *Economics Planning of Super Tall Buildings in Asia Pacific Cities*, Strategic Integration of Surveying Services FIG Working Week 2007, Hong Kong, China, 13–17 May 2007.

Holmes, J.D., *Wind Loading of Structures*, Spon Press, London, 2001.

Huxtable, A.L., *The Tall Building Artistically Reconsidered: The Search for a Skyscraper Style*, Pantheon Books, Random House Inc., New York, 1984.

Irwin, P.A., *The Wind Engineering of the Burj Dubai Tower*, CTBUH 7th World Congress 2005, New York, 2005.

Irwin, P.A., Bluff Body Aerodynamics in Wind Engineering, *Journal of Wind Engineering and Industrial Aerodynamics*, Vol. 96, pp. 701–712, 2008.

Irwin, P.A., Wind Engineering Challenges of the New Generation of Super-Tall Buildings, *Journal of Wind Engineering and Industrial Aerodynamics*, Vol. 97, pp. 328–334, 2009.

Irwin, P.A. and Baker, W.F., The Burj Dubai Tower: Wind Engineering, *Structure Magazine*, pp. 28–31, June 2006.

Irwin, P.A., Baker, W.F., Korista, S., Weismantle, P.A. and Novak, L.C., The Burj Dubai Tower: Wind Tunnel Testing of Cladding and Pedestrian Level, *Structure Magazine*, pp. 48–51, November 2006.

Irwin, P.A., Kilpatrick, J. and Frisque, A., *Friend or Foe, Wind at Height*, CTBUH 8th World Congress 2008, Dubai, 2008a.

Irwin, P.A., Kilpatrick, J., Robinson, J. and Frisque, A., Wind and Tall Buildings: Negatives and Positives, *The Structural Design of Tall and Special Buildings*, Vol. 17, pp. 915–928, 2008b.

Irwin, P.A., Denoon, R. and Scott, D., *Wind Tunnel Testing of High-rise Buildings*, CTBUH in conjunction with ITT and Routledge/Taylor and Francis Group, Chicago, 2013.

Isyumov, N., Fediw, A., Colaco, J. and Banavalkar, P.V., Performance of a Tall Building Under Wind Excitation, *Journal of Wind Engineering and Industrial Aerodynamics*, Vol. 41–44, 1992.

Iyengar, H., Reflections on the Hancock Concept, *CTBUH Review*, Vol. 1, p. 1, May 2000.

Jodidio, P., *Renzo Piano Building Workshop: 1966 to Today*, Taschen, 2008.

Joseph, M.L., Poon, D.C.K. and Shieh, S., Taipei 101: The World's Tallest Building, *Structure Magazine*, pp. 40–45, June 2006.

Kareem, A., Lateral-torsional Motion of Tall Buildings to Wind Loads, *Journal of Structural Engieering*, ASCE, Vol. 111, No. 11, pp. 2479–2496, 1985.

Kareem, A., Kijewski, T. and Tamura, Y., Mitigation of Motion of Tall Buildings with Specific Examples of Recent Applications, *Wind and Structures*, Vol. 2, No. 3, pp. 201–251, 1999.

Katz, P. and Robertson L.E., Case Study: Shanghai World Financial Center, *CTBUH Journal*, No. 2, 2008.

Katz, P., Robertson, L.E. and See, S., *Shanghai World Financial Center: Without Compromise*, CTBUH 8th World Congress 2008, Dubai, 2008.

Kawai, H., Effect of Corner Modifications on Aeroelastic Instabilities of Tall Buildings, *Journal of Wind Engineering and Industrial Aerodynamics*, Vol. 74–76, pp. 719–729, 1998.

Khan, F.R., *Recent Structural Systems in Steel for High-Rise Buildings*, Proceedings of the British Constructional Steelwork Association Conference on Steel in Architecture. London: British Constructional Steelwork Association, 1969.
Khan, F.R., *The Future of High Rise Structures*, Progressive Architecture, 1972.
Khan, F.R., Evolution of Structural Systems for High-Rise Buildings in Steel and Concrete, in J. Kozak (Ed.), *Tall Buildings in the Middle and East Europe: Proceedings of the 10*. Regional Conference on Tall Buildings-Planning, Design and Construction, 1973.
Kidder Smith, G. E., *The New Architecture of Europe*, Pelican Books, pp. 179–180, 1962.
Kijewski-Correa, T., *Shaping the Wing and the Face of Tomorrow's Skyscrapers*, A Final Report to the SOM Foundation, Vol. 1, 2002.
Kikitsu, H. and Okada, H., *Characteristics of Aerodynamic Response of High-rise Buildings with Open Passage*, Proceedings of the CIB-CTBUH International Conference on Tall Buildings, Malaysia, 2003.
Kim, H. and Elnimeiri, M., *Space Efficiency in Multi-Use Tall Building*, CTBUH Seoul Conference, pp. 271–278, Seoul, 2004.
Kim, J.O. and Lee, S.H., *Application of High Performance Concrete in Petronas Twin Tower, KLCC*, CTBUH Seoul Conference, pp. 457–464, Seoul, 2004.
Kim, Y., You, K. and Ko, N., Across-wind Response of an Aeroelastic Tapered Tall Buildings, *Journal of Wind Engineering and Industrial Aerodynamics*, Vol. 96, pp. 1307–1319, 2008.
Koolhaas, R. and Scheeren, O., *CCTV by OMA*, A+U Co., Ltd., Japan, July 2005.
Kwok, K.C.S., Cross-wind Response of Tall Buildings, *Engineering Structures*, Vol. 4, 1982.
Kwok, K.C.S., *Aerodynamics of Tall Buildings*, A State of the Art in Wind Engineering: International Association for Wind Engineering, 9th International Conference on Wind Engineering, New Delhi, 1995.
Kwok, K.C.S. and Bailey, P.A., Aerodynamic Devices for Tall Buildings and Structures, *ASCE Journal of Engineering Mechanics* 113, pp. 349–365, 1987.
La Torre Pirelli a Milano, Domus, No. 348, pp. 1–3, 1958.
Lew, H.S. and Sadek, F., *WTC Towers: Innovative Design Features and Structural Modeling*, CTBUH Seoul Conference, pp. 1150–1153, Seoul, 2004.
Lu, X., Zou, Y., Lu, W. and Zhao, B., Shaking Table Model Test on Shanghai World Financial Center Tower, *Earthquake Engineering and Structural Dynamics*, Vol. 36, pp. 439–457, 2007.
Lubell, S., Sports City Tower, Doha, Qatar, *Architectural Record*, August 2007.
Martin, O. and Dean, B., *Advances in the Structural Design of High-Rise Residential Buildings in Australia*, CTBUH 7th World Congress 2005, New York, 2005.
Millais, M., *Building Structures: From Concepts to Design*, Taylor and Francis Books, 2005.
Miyashita, K., Nakamura, O., Tagaya, K. and Abiru, H., *Wind Tunnel Tests and Full Scale Measurements of High-Rise Building Equipped with TAD*, Proceedings of the 5th East Asia-Pacific Conference on Structural Engineering and Construction, pp. 1262–1266, 1995.
Mufti, A.A. and Bakht, B., Fazlur Khan (1929–1982): Reflections on His Life and Works, *Canadian Journal of Civil Engineering*, Vol. 29, pp. 238–245, 2002.
Naidu, K.G., *The Petronas Tower: The World's Tallest Building*, CPAC Seminar 95, Modern Technology in Concrete Construction, Bangkok, Thailand, 1995.
Peet, G., The Origin of the Skyscraper, *CTBUH Journal*, Issue I, pp. 18–23, 2011.
Pelli, C., Skyscrapers, *Perspecta, Yale University Journal*, Vol. 18, pp. 134–147, 1982.
Pollalis, S., *The Petronas Towers*, Harvard Design School, Center for Design Informatics, 2002.
Ponti, G., *A specific letter in which Gio Ponti explained his design philosophy was sent to Prof. Enis Kortan*, 27 October 1977.
Ponti, G., Espressione, dell'edificio Pirelli in costruzione a Milano, *Domus*, No. 316, pp. 1–16, 1956.
Ponti, G., Immeuble Pirelli, Milan, *L'Architecture d'Aujourd'hui*, No. 64, pp. 2–5, 1956.
Ponti, G., Out of Philosophy of Architecture, *Architectural Record*, pp. 155–164, 1956.

Poon, D.C.K., Shieh, S., Joseph, M.L. and Chang, C., Reaching for the Sky, *Civil Engineering*, pp. 56–62, January 2004.
Poon, D.C.K., Shieh, S., Joseph, M.L. and Chang, C., *Structural Design of Taipei 101: The World's Tallest Building*, CTBUH Seoul Conference, pp. 271–278, Seoul, 2004.
Reina, P., *Architect-Engineer Santiago Calatrava Deftly 'Twists' Tower*, ENR, December 13, 2004.
Robertson, L.E. and See, S., The Shanghai World Financial Center, *Structure Magazine*, pp. 32–35, June 2007.
Scarangello, T.Z., Joseph, L.M. and Krall, K.E., A Statement in Steel: The New York Times Building, *Structural Engineer*, June 2004.
Scarangello, T.Z., Krall, K.E. and Callow, J.A., *A Statement in Steel: The New York Times Building*, CTBUH 8th World Congress 2008, Dubai, 2008.
Schueller, W., *High-Rise Building Structures*, John and Wiley Sons Inc., New York, 1977.
Scott, D., Hamilton, N. and Ko, E., Structural Design Challenges for Tall Buildings, *Structure Magazine*, Vol. 2, pp. 20–23, February 2005.
Smith, A., *Burj Dubai: Designing the World's Tallest*, UAE, CTBUH 8th World Congress 2008, Dubai, 2008.
Smith, B.S. and Coull, A., *Tall Building Structures: Analysis and Design*, New York: Wiley, 1991.
Smith, R.F. and Killa, S., Bahrain World Trade Center (BWTC): The First Large-Scale Integration of Wind Turbines in a Building, *The Structural Design of Tall and Special Buildings*, Vol. 16, pp. 429–439, 2007.
Taranath, B., *Steel, Concrete & Composite Design of Tall Buildings*, New York: McGraw-Hill Book Company, 1998.
Taranath, B., *Wind and Earthquake Resistant Buildings: Structural Analysis and Design*, A Series of Reference Books and Textbooks (Editor: Michael D. Meyer), Department of Civil and Environmental Engineering, Georgia Institute of Technology, Atlanta, Georgia, 2005.
The Architectural Review, Eureka Tower, Melbourne, Australia Fender Katsalidis Architects, No. 1340, October 2008.
Thornton, C., Hungspruke, U. and Joseph, L., Design of the World's Tallest Buildings – Petronas Twin Towers at Kuala Lumpur City Center, *The Structural Design of Tall Buildings*, Vol. 6, pp. 9–27, 1997.
Tomasetti, T., *The Chicago Spire*, Thornton Tomasetti Building Solutions Annual Report, 2007.
Turning Torso, *Architectural Record*, Building Types Study – Tall buildings, August 2007.
Turning Torso, Yapı Dergisi, No. 290, pp. 54–59, January 2009. (Translated from English by Burçin Yılmaz)
Weingardt, R.G., Great Achievements – Notable Structural Engineers: Fazlur Rahman Khan, The Einstein of Structural Engineering, *Structure Magazine*, February 2011.
Weismantle, P.A., Smith, G.L. and Sheriff, M., Burj Dubai: An Architectural Technical Design Case Study, *The Structural Design of Tall and Special Buildings*, Vol. 16, pp. 335–360, 2007.
WS Atkins & Partners, *Bahrain World Trade Center: The World's First Large Scale Integration of Wind Turbines on a Building, CTBUH Best Tall Building Awards*, pp. 1–21, WS Atkins & Partners Publication, 2008.
Zaknic, I., Smith, M. and Rice, D., *100 of The World's Tallest Buildings*, Council on Tall Buildings and Urban Habitat Committee, New York: McGraw-Hill Book Company, 1998.
Ziegler, C., Out of Ashes and Rubble: The Pirelli Tower, *Places*, College of Environmental Design, UC Berkeley, Vol. 21, Issue: 1, 2009.

INDEX